Latent Trait and Latent Class Models

Latent Trait and Latent Class Models

Edited by

Rolf Langeheine

and

Jürgen Rost

Institute for Science Education at the
University of Kiel
Kiel, Federal Republic of Germany

Plenum Press • New York and London

Library of Congress Cataloging in Publication Data

Latent trait and latent class models / edited by Rolf Langeheine and Jürgen Rost.
   p.   cm.
   Includes bibliographical references and index.
   ISBN 0-306-42727-3
   1. Latent structure analysis. I. Langeheine, Rolf. II. Rost, Jürgen. III. Title: Latent
class models.
QA278.6.L33   1988
519.5′35 — dc19

88-4141
CIP

© 1988 Plenum Press, New York
A Division of Plenum Publishing Corporation
233 Spring Street, New York, N.Y. 10013

Printed in the United States of America

Contributors

ERLING B. ANDERSEN, Department of Statistics, University of Copenhagen, DK-1455 Copenhagen K, Denmark

GERHARD ARMINGER, Department of Economics, University of Wuppertal, D-5600 Wuppertal 1, Federal Republic of Germany

JOHN R. BERGAN, Program in Educational Psychology, College of Education, University of Arizona, Tucson, Arizona 85721

CLIFFORD C. CLOGG, Department of Statistics, Pennsylvania State University, University Park, Pennsylvania 16802

C. MITCHELL DAYTON, Department of Measurement, Statistics, and Evaluation, College of Education, University of Maryland, College Park, Maryland 20742

G. DUNN, Biometrics Unit, Institute of Psychiatry, University of London, London SE5 8AF, United Kingdom

B. S. EVERITT, Biometrics Unit, Institute of Psychiatry, University of London, London SE5 8AF, United Kingdom

FRANK KOK, Psychological Laboratory, University of Amsterdam, 1018 XA Amsterdam, The Netherlands

KLAUS D. KUBINGER, Institute of Psychology, University of Vienna, A-1010 Vienna, Austria

ULRICH KÜSTERS, Department of Economics, University of Wuppertal, D-5600 Wuppertal 1, Federal Republic of Germany

Rolf Langeheine, Institute for Science Education (IPN), University of Kiel, D-2300 Kiel 1, Federal Republic of Germany

George B. Macready, Department of Measurement, Statistics, and Evaluation, College of Education, University of Maryland, College Park, Maryland 20742

Geofferey Norman Masters, Centre for the Study of Higher Education, University of Melbourne, Parkville, Victoria 3052, Australia

Jürgen Rost, Institute for Science Education (IPN), University of Kiel, D-2300 Kiel 1, Federal Republic of Germany

Fons J. R. van de Vijver, Department of Psychology, Tilburg University, 5000 LE Tilburg, The Netherlands

Arnold L. van den Wollenberg, Department of Mathematical Psychology, University of Nijmegen, 6500 HE Nijmegen, The Netherlands

Preface

This volume is based on an international conference held at the Institute for Science Education (IPN) in Kiel in August 1985. The IPN is a national research institute for science education of the Federal Republic of Germany associated with the University of Kiel.

The aim of this conference—to treat latent trait and latent class models under comparative points of view as well as under application aspects—was realized in many stimulating contributions and very different ways. We asked the authors of these papers to work out their contributions for publication here, not only because many of the papers present new material, but also because the time is ripe for a comprehensive volume, working up the widespread literature of the past ten years in this field.

We have tried to compile a volume that will be of interest to statistically oriented researchers in a variety of disciplines, including psychology, sociology, education, political science, epidemiology, and the like. Although the chapters assume a reasonably high level of methodological sophistication, we hope that the book will find its way into advanced courses in the above fields.

We are grateful to the IPN for organizing the conference, to our contributors for their untiring efforts in revising their chapters for publication, and to the staff of Plenum Publishing Corporation for helping to make this book a reality.

Contents

PART I LATENT TRAIT THEORY

PART II   LATENT CLASS THEORY

CHAPTER 4

New Developments in Latent Class Theory   77

ROLF LANGEHEINE

CHAPTER 5

Log-Linear Modeling, Latent Class Analysis, or Correspondence
Analysis: Which Method Should Be Used for the Analysis of
Categorical Data? 109

B. S. EVERITT AND G. DUNN

CHAPTER 6

A Latent Class Covariate Model with Applications to
Criterion-Referenced Testing 129

C. MITCHELL DAYTON AND GEORGE B. MACREADY

PART III   COMPARATIVE VIEWS OF LATENT TRAITS AND
LATENT CLASSES

CHAPTER 7

Test Theory with Qualitative and Quantitative Latent Variables . . 147

JÜRGEN ROST

PART IV   APPLICATION STUDIES

CHAPTER 10

Latent Variable Techniques for Measuring Development 233

JOHN R. BERGAN

CHAPTER 11

Item Bias and Test Multidimensionality 263

FRANK KOK

CHAPTER 12

On a Rasch-Model-Based Test for Noncomputerized Adaptive
Testing 277

KLAUS D. KUBINGER

CHAPTER 13

Systematizing the Item Content in Test Design 291

FONS J. R. VAN DE VIJVER

Introduction and Overview

ROLF LANGEHEINE AND JÜRGEN ROST

1. INTRODUCTION

The primary goal of this volume is to present a synthesis of two research areas which for a long time developed independently. Latent trait models came from the tradition of mental test theory and were primarily applied in psychological and educational studies. Latent class models, in contrast, were developed in the context of cross-table analysis and log-linear modeling and were mostly applied to sociological research data. This separation has taken place despite the fact that there had been connections between the two at an early stage. For instance, the most famous book in this field (*Latent Structure Analysis* by Lazarsfeld & Henry, 1968) treated both topics within a common framework, and some of the ideas of these authors are taken up and advanced in the present volume.

The aim of bringing together latent trait and latent class theory is for each research area to profit from developments in the other. One example of how this may work involves testing the fit of the different kinds of models. Whereas a variety of goodness-of-fit tests has been developed for Rasch models (cf. the contribution by van den Wollenberg), latent class analysis has essentially only one goodness-of-fit test, presenting some problems when the number of items and categories is large. Here, latent class theory could benefit from Rasch's theory. On the other hand, the only model test used for latent class models has been largely unknown to researchers on Rasch's theory. In fact, the refor-

ROLF LANGEHEINE AND JÜRGEN ROST • Institute for Science Education (IPN), University of Kiel, D-2300 Kiel 1, Federal Republic of Germany.

mulation of the Rasch model as a log-linear model shows further ways of testing the model (cf. the contribution by Clogg). Another example of mutual benefit is the transfer of the threshold concept in latent trait theory to latent class models (cf. the contribution by Rost). The chapters of the present volume give many further instances of such an influence.

A synthesis of two research areas can be tried for various reasons and from different points of view. For instance, one can stress the common theoretical concepts and contrast them with the differences between the two approaches. Another aspect would be to search for a supertheory subsuming both approaches as submodels. Or someone may show that analyses usually done with one type of modeling can also be done by the other one and vice versa. Furthermore, both theories can be treated as standing in competition, fighting for the better fit to the same sets of data. A more constructive point of view would be to transfer concepts and ideas from one theory to the other, for example, using the same restrictions or making the same extensions of the basic models.

The present book deals with all of these angles and the following overview with cross-references to the chapters of the book serves as a guideline for those readers who have special interests in one or more of the above issues. The reader who is looking for an *introduction* to the field is referred to the contribution of Masters regarding latent trait theory, and to that of Langeheine for latent class theory. Those who can best understand a model via its *applications* to real data are on the one hand referred to Part IV of the volume, where for example, Kubinger describes the application of the Rasch model to a *tailored testing* problem and Kok outlines a theory of *item bias*. On the other hand each of the papers in Parts II and III also provides applications of the models discussed there to real data examples.

2. COMMON CONCEPTS OF LATENT TRAIT AND LATENT CLASS MODELS

The contributions of Andersen and Clogg describe in detail the most fundamental feature of both types of models, namely the construction of a latent variable making the observed variables mutually independent, if that latent variable is conditioned out. Needless to say, the latent variable is quantitative and continuous in the one case, and categorical-discrete in the other case. This common property is called *local independence,* because independence of the observed variables only exists when the 'locus' of all individuals on the latent variable is the same, that is, is held constant. Local independence may be treated as an *axiom* or an unrealistic but necessary *assumption*. The contribution by Andersen, however, makes clear that it is the *goal* of any latent structure analysis and it guides us to find a latent variable explaining all observed dependencies.

There are other common concepts in both theories which are of a less fundamental nature. Those topics will be mentioned in the course of the following paragraphs.

3. THE SEARCH FOR THE MOST GENERAL MODEL

When there are two theories having fundamental features in common the question arises whether one of the two is the more general one covering the other as a submodel, or if neither, what a joint supermodel looks like. Hierarchies of more or less general models can be construed within each approach. Masters shows such a hierarchy of polytomous Rasch models in his contribution and Langeheine investigates the question whether one of the different formulations of latent class analysis—Goodman's, Formann's or Haberman's—is the most general one. The answer to the question for the joint generalization of latent trait and latent class models, however, can most easily be found in the contribution by Clogg. He points out how a Rasch model can be formalized as a log-linear model. From Haberman's model we know that latent class analysis is also a log-linear model, but extended by a latent variable. Hence, the *log-linear model structure* with latent variables is the joint supermodel of latent trait and latent class analysis.

Specifically, it turns out that the Rasch model formalized as a log-linear model needs no latent variable, whereas all unrestricted latent class models do. From this it seems as if the latent class approach is the more general one and the contribution by Clogg demonstrates impressively that it is also the more flexible. However, the possibility of representing the Rasch model without any latent variable is a consequence of its favorable property of "conditioning out" the person variable, that is, to allow item parameter estimation without knowing the person parameters and their distribution.

4. ORDERED CLASSES AND DISCRETE TRAITS

The conclusion that latent class analysis is more general than latent trait analysis seems to turn things upside down when proceeding from the concept that latent classes do represent a discrete distribution along a latent continuum. From that point of view, a distribution of individuals into latent classes is a special type of distribution of a latent variable and, hence, a special case of a latent trait model allowing any kind of distribution. Of course, this view of things only holds if the latent classes can be ordered along a latent continuum.

The concept of ordered classes has already been treated by Lazarsfeld and Henry (1968) and is advanced in several contributions of the present volume. According to this concept, the order of latent classes can be defined in terms of a uniform order of all response probabilities across the classes, that is, the rank of a latent class is higher than that of another class, if *all* response probabilities of that class are higher than in the other class. Such ordered classes can in principle be regarded as representing a discrete distribution of a continuous latent variable. Hence, ordered classes and discrete traits are two sides of the same coin. However, those models that are at the same time latent class and latent trait models have many facets.

In one case, the empirical distribution is in fact a discrete distribution, that is, the latent variable takes on only a restricted number of values within the population under consideration. Latent class analysis then produces classes which are located at distinct points of a continuum, or in the terminology of Lazarsfeld and Henry *located classes*. Clogg deals in his contribution with the situation of located classes in the case of Rasch-scaled items, a model first introduced by Formann (1985). Rost treats the unrestricted case of ordered located classes, which can be produced by any items with monotone item characteristic curves. An extreme case of located classes are the two classes of "masters" and "nonmasters" in the *mastery-testing* situation. This is dealt with in the papers by Dayton and Macready and Bergan.

Another case of coming to a discrete distribution has already been mentioned by Lazarsfeld and Henry.

> Sometimes, of course, a discrete distribution is an approximation to a continuous distribution. The "classes" are then simply artifacts, a means of summarizing some of the information about the population. In latent structure analysis, however, the requirement that the classes be homogeneous makes such an interpretation invalid, for it is very hard to find good approximations to local independence. (1968, pp. 148–149)

With regard to the latter qualification, Clogg's paper goes a step further. He shows that the Rasch model is equivalent to a restricted latent class model having as many classes as there are different person parameter estimates, namely, the number of items minus one.

A third case dealing with a discrete distribution and hence with ordered classes is the area of *latent distance* models, *scaling* models, *response error* models or generally the latent trait models whose item characteristic curves are step-functions. In this case, the distribution of the latent variable—being of any shape—can be treated as a discrete distribution, because all individuals located between two neighbouring steps on the latent continuum may be grouped into the same latent class. The papers of Langeheine and Clogg deal with these models in detail. The latter author, in particular, treats the problem of *scaling* the classes obtained in such an analysis.

5. THE USE OF RESTRICTIONS FOR EXTENDING APPLICABILITY

The specification and application of a latent distance model within the framework of latent class analysis is not a matter of extending but of restricting the basic latent class model. This may sound paradoxical, but latent class theory is full of examples where restrictions were used to extend applicability. The contribution by Langeheine gives an impressive example of what can be done using only fixed parameters and *equality constraints*. Bergan gives examples for the same kind of restrictions within the latent trait framework, which have hardly ever been used in earlier studies. Andersen introduces a restriction, which is very unusual for the "sample-free" Rasch model, but very popular in other fields: the restriction of a *normally distributed latent variable*. These latter examples are aimed less at extending the applicability of the models than at saving parameters and testing special hypotheses, two further relevant functions of model restrictions.

The latent class models for *ordered categories* introduced in the paper by Rost are also generated by restricting the basic model. Through the use of this kind of threshold restriction the applicability of latent class analysis has been extended to manifest variables with ordered categories, a field that could hardly be dealt with using only equality constraints and fixed parameters. These restrictions are in fact *linear decompositions* of exponential parameters in a logistic model structure and, hence, of the same nature as the linear logistic latent class model (cf. the contribution of Langeheine) and the linear logistic Rasch model (cf. the contribution by van de Vijver).

Finally, Langeheine, discusses models for the analysis of *sequential categorical data* using restrictions with respect to Haberman's log-linear formulation of latent class analysis.

6. EXTENSIONS OF THE BASIC MODELS

Besides those extensions of model applicability described above, real extensions of the *model structure* are presented in some of the contributions. Arminger and Küsters describe an integration of Bock's (1972) multinomial logit latent trait model into what is best known under the heading of LISREL structure. Hence, not only models with more than one latent variable are covered but also manifest variables of different measurement levels. Another approach presented by Andersen introduces more than one latent variable. He formulates and applies a Rasch model with two latent variables, which includes the latent correlation between these two normally distributed variables as an additional model parameter.

In contrast to these extended latent trait models, the construction of latent class models with more than one latent (categorical) variable can be done by means of parameter restrictions. The contributions by Andersen, Bergan and Langeheine treat this topic in detail.

Dayton and Macready present quite a different extension of standard latent class analysis, namely, the introduction of a manifest continuous covariate variable of latent class membership. This may be a first step towards more complex latent class models, like those proposed by Arminger and Küsters for metric latent variables.

7. TESTING AND COMPARING THE FIT OF MODELS

Undoubtedly, the existence of exact goodness-of-fit tests is one of the predominant features of latent trait and latent class models. As mentioned earlier, the variety of model tests for Rasch models is very large (cf. van den Wollenberg) compared with the Pearson chi-square and the likelihood ratio statistic in latent class theory. But in any case it is possible to test the null hypothesis, stating that a specific model holds true for the data.

Such a model test, however, only shows that the model under consideration is one model fitting the data and says nothing about other models that may fit the same data equally well or even better. The area of comparing the fit of alternative models and the contributions of this volume give many examples of testing one model against the other.

If the models to be compared are related hierarchically, that is, one model is a submodel of the other, exact tests in the form of the likelihood ratio statistic are available. Deciding between a latent trait and a latent class model is usually another matter, because they are not hierarchically related. Andersen makes comparisons by evaluating the chi-square values of both models taking the number of estimated parameters into account. Rost also gives examples of comparing latent trait and latent class models by means of the likelihood of data under different models as well as by means of chi-square statistics. Both papers deal with the problem of significance tests in sparse tables and propose pattern grouping to exceed a minimum value of expected pattern frequencies.

Yet another way of comparing different models is to analyze data according to these methods and to focus on what we can learn from the results. Everitt and Dunn thus show that the applied researcher may gain different insights into his/her data depending on whether a log-linear, latent class or a correspondence analysis is made. Whereas those relying on formal hypothesis testing would prefer the log-linear or latent class approach, promotors of correspondence analysis would argue against "defective inferential methods based on questionable assumptions" and instead would recommend simply looking at graphical displays produced by, for example, correspondence analysis "in the hope of discovering an interesting pattern or structure." It can be shown, however—and Everitt and Dunn comment on this—that both latent class and correspondence analysis can be expressed in log-linear form. Log-linear and correspondence analysis may thus be used in a complementary form as advocated by, for example, Van der Heijden and de Leeuw (1985), combining the merits of both approaches (i.e., geometrical representation and formal testing). See as well the recent paper by Goodman (1985) for the further development of inferential aspects of correspondence analysis.

The number of different topics treated in the contributions of the present book shows that the process of integrating latent trait and latent class theory has not yet been completed. But the contributions also show that the attempt at such an integration is a fruitful enterprise.

8. REFERENCES

Bock, R. D. (1972). Estimating item parameters and latent ability when responses are scored in two or more nominal categories. *Psychometrika, 37*, 29–51.

Formann, A. K. (1985). Constrained latent class models: Theory and applications. *British Journal of Mathematical and Statistical Psychology, 38*, 87–111.

Goodman, L. A. (1985). The analysis of cross-classified data having ordered and/or unordered categories: Association models, correlation models, and asymmetry models for contingency tables with or without missing entries. *Annals of Statistics, 13*, 10–69.

Lazarsfeld, P. F., & Henry, N. W. (1968). *Latent structure analysis*. Boston: Houghton Mifflin.

Van der Heijden, P. G. M., & de Leeuw, J. (1985). Correspondence analysis used complementary to loglinear analysis. *Psychometrika, 50*, 429–447.

PART I

LATENT TRAIT THEORY

CHAPTER ONE

Measurement Models for Ordered Response Categories

GEOFFEREY NORMAN MASTERS

1. INTRODUCTION

Quantitative educational research depends on the availability of carefully constructed variables. The construction and use of a variable begin with the idea of a single dimension or line on which students can be compared and along which progress can be monitored. This idea is operationalized by inventing items intended as indicators of this latent variable and using these items to elicit observations from which students' positions on the variable might be inferred.

In the construction of educational variables, individual test or questionnaire items are used only as instances of the variable that we hope to define. An item like $34 \times 29 = ?$ from a mathematics achievement test, for example, serves only as an instrument for the collection of observations from which measures of mathematics achievement might be inferred. It is our intention that such items should be interchangeable: the item $34 \times 29 = ?$ should be replaceable with other items of a similar type (perhaps $28 \times 67 = ?$ or $77 \times 23 = ?$). Comparability of measures made with different test items is a fundamental requirement in the practice of educational measurement. Within some specified context, measures of student attainment must be comparable across different choices of test items.

This intention can be illustrated by considering two persons m and n

GEOFFEREY NORMAN MASTERS • Centre for the Study of Higher Education, University of Melbourne, Parkville, Victoria, 3052, Australia.

FIGURE 1. Locations β_m and β_n of persons m and n on variable X.

with locations β_m and β_n on some latent variable X (see Figure 1). Persons m and n can be compared on this variable by having them take the same set of items and then using their responses to these items to estimate the difference $\beta_n - \beta_m$ between their locations. The intention of interchangeability is the intention that, within some useful set of items, this estimated difference will be invariant with respect to the choice of items.

The estimated difference $\hat{\beta}_n - \hat{\beta}_m$ can be freed of the choice of items only if all items in a set concur on the difference between persons m and n. An item that works against the other items to make persons m and n appear closer together or further apart on this variable spoils the opportunity for item-free comparison.

In educational measurement, an item can work against the other items in a set for a variety of reasons. Flawed items that introduce ambiguity through their wording, or which are easy to get right for the wrong reason (e.g., items for which it is unusually easy to guess the right answer), or which have no correct answer, or which have more than one correct answer, introduce noise and so reduce apparent differences on the variable of interest.

Items sensitive to differences on a second, contaminating, variable also work to reduce the apparent differences between individuals. An item on a reading comprehension test that depends on background knowledge possessed by some students but not by others introduces a bias that favors particular students. Because some poorer readers have a special advantage on this item, and some better readers have a special disadvantage, the general consequence of this type of disturbance is to make students look more similar in reading ability than they would in the absence of this item.

An item can also work against other items to *increase* the apparent difference between some students. This is a special type of measurement disturbance of particular concern in the construction of educational variables. It arises most commonly when test items interact with differences in school curricula. Students streamed into different instructional levels frequently take the same achievement tests. However, items on these tests sometimes cover content that has been taught to students in one stream but not to students in the other. An item that covers content taught only to students in an advanced stream gives those students a special advantage over students in lower streams who have not covered that content. The result is to make students in high and low streams appear more different than they would if all items were based on content that all students had covered.

For the construction of educational variables, a measurement model must meet at least three requirements. It must provide a way of estimating from observed responses to test and questionnaire items the locations of individuals on the latent dimension that these items are intended to define. It must provide a way of freeing these estimates from the choice of items so that measures have a generality that extends beyond the particular items used. And it must provide a framework for the identification of items that introduce contaminating influences like those outlined above.

2. OBJECTIVE COMPARISON

The key to freeing the comparison of two person parameters β_m and β_n from the particular choice of items used to make this comparison was developed by Georg Rasch in the 1950s in his work on measuring reading proficiency. To illustrate how this type of comparison is made possible, we begin with Rasch's probabilistic model for the success of person m on dichotomously scored test item i:

$$\pi_{mi} = \exp(\beta_m - \delta_i)/[1 + \exp(\beta_m - \delta_i)] \tag{1}$$

where π_{mi} is the probability of person m succeeding on item i, β_m is the ability of person m, and δ_i is the difficulty of item i. The model probability of a second person n succeeding on this same item i is

$$\pi_{ni} = \exp(\beta_n - \delta_i)/[(1 + \exp(\beta_n - \delta_i)] \tag{2}$$

When these two persons m and n attempt item i, there are four possible outcomes: both fail $(0, 0)$, person m succeeds but n fails $(1, 0)$, person m fails but n succeeds $(0, 1)$, and both succeed $(1, 1)$. Only two of these four possible outcomes $(1, 0$ and $0, 1)$ provide information about the relative locations of persons m and n on this variable. Assuming local independence of the responses of persons m and n to item i, the probabilities of these two outcomes are

$$\pi_{10i} = \pi_{ni}(1 - \pi_{mi}) = \exp(\beta_n - \delta_i)/\Psi \tag{3}$$

and

$$\pi_{01i} = (1 - \pi_{ni})\pi_{mi} = \exp(\beta_m - \delta_i)/\Psi \qquad (4)$$

where π_{10i} is the probability of person m succeeding on item i but person n failing, π_{01i} is the probability of person m failing but person n succeeding, and $\Psi = [1 + \exp(\beta_n - \delta_i)][1 + \exp(\beta_m - \delta_i)]$.

From (3) and (4) it follows that the conditional probability of outcome $(1, 0)$, given that one of these two persons fails item i and the other succeeds, is

$$\frac{\pi_{10i}}{\pi_{01i} + \pi_{10i}} = \frac{\exp(\beta_n - \beta_m)}{1 + \exp(\beta_n - \beta_m)} \qquad (5)$$

The significance of (5) is that it involves only the person parameters β_n and β_m and does not depend at all on the item parameter δ_i. The implications of this can be seen by rewriting (5) as

$$\beta_n - \beta_m = \ln(\pi_{10i}/\pi_{01i}) \qquad (6)$$

from which it follows that an estimate of the difference $\beta_n - \beta_m$ is given by

$$\hat{\beta}_n - \hat{\beta}_m = \ln(N_{10}/N_{01}) \qquad (7)$$

where N_{10} is the number of items answered correctly by person m but failed by person n, and N_{01} is the number of items failed by person m but answered correctly by person n, regardless of the difficulties of those items. When test data conform to (1), it is possible to construct a conditional probability (5) which frees the comparison of β_m and β_n from the particulars of the items used.

Equation (7) also suggests a simple test of the extent to which different subsets of items point to the same estimate of $\beta_n - \beta_m$. By obtaining the counts N_{10} and N_{01} separately for different subsets of items (e.g., addition and subtraction items), separate estimates of the $\beta_n - \beta_m$ difference can be obtained and compared directly.

Consider now the comparison of two test items i and j with locations δ_i and δ_j on variable X (see Figure 2). From (1) the model probabilities of a particular person n succeeding on each of items i and j can be obtained. Assuming local independence of person n's attempts at these two items, expressions for the probability π_{n10} of person n passing item i but failing item j, and the probability π_{n01} of this same person failing item

FIGURE 2. Locations δ_i and δ_j of items i and j on variable X.

i but passing item j can be obtained and used to construct the conditional probability

$$\frac{\pi_{n10}}{\pi_{n01} + \pi_{n10}} = \frac{\exp(\delta_j - \delta_i)}{1 + \exp(\delta_j - \delta_i)} \qquad (8)$$

which contains only the difficulties of the two items being compared. The difference between items i and j on variable X can be estimated as

$$\hat{\delta}_j - \hat{\delta}_i = \ln(N_{10}/N_{01}) \qquad (9)$$

where N_{10} is the number of persons getting i right but j wrong, and N_{01} is the number of persons getting i wrong but j right, regardless of the abilities of those persons.

The crucial feature of Rasch's dichotomous model (1) for objective comparison is that it enables the construction of the conditional probabilities (5) and (8) from which all parameters but the pair being compared are eliminated. In (5) and (8) the difference between these two parameters is linked to two observable events. These are two of several possible observable events, but they are the only two events containing information about the relative values of these two parameters.

In retrospect, (1) can also be thought of as a conditional probability of the same form as (5) and (8). In (1), the local comparison is between the two parameters β_n and δ_i (see Figure 3). Although (1) is not usually described as a conditional probability, it can be represented as such if we recall that the observed outcomes are constrained to only two possibilities (e.g., right and wrong). Other outcomes like failure to respond are not covered by (1), which gives the probability of a correct response conditional on an answer being scored either right or wrong. With this observation, (1) can be rewritten

$$\frac{\pi_{ni1}}{\pi_{ni0} + \pi_{ni1}} = \frac{\exp(\beta_n - \delta_i)}{1 + \exp(\beta_n - \delta_i)} \qquad (10)$$

FIGURE 3. Locations β_n and δ_i of person n and item i on variable X.

which is clearly of the same form as (5) and (8), and contains only the two parameters β_n and δ_i being compared. The two observable events in (10) are success of person n on item i and failure of person n on item i. The probability of person n succeeding on item i given that this person either succeeds or fails is governed only by the relative values of β_n and δ_i. If it were possible for person n to interact with item i a number of times such that the outcome of one interaction did not influence any other interaction, then the $\beta_n - \delta_i$ difference could be estimated as

$$\hat{\beta}_n - \hat{\delta}_i = \ln(N_1/N_0) \tag{11}$$

where N_1 is the number of times person n succeeds on item i, and N_0 is the number of times person n fails item i. Once again, the difference between the two parameters being compared can be estimated from counts of two observable events.

The identical forms of (5), (8), and (10) invite the following generalization: *Two parameters ξ_j and ξ_k can be compared objectively (i.e., independently of the particulars of the observations), within a particular frame of reference, if there exist two mutually exclusive observable events A and B and, within this frame of reference, the probability of any particular observation i being event B rather than event A is governed only by*

$$\frac{\pi_{Bi}}{\pi_{Ai} + \pi_{Bi}} = \frac{\exp(\xi_k - \xi_j)}{1 + \exp(\xi_k - \xi_j)} \tag{12}$$

where *π_{Ai} is the probability of observation i being event A, and π_{Bi} is the probability of this same observation being event B.*

Equation (12) is Rasch's elementary model for objective comparison. When data approximate this model, the estimated difference $\hat{\xi}_k - \hat{\xi}_j$ can be freed of the particulars of the observations used to make this comparison. The possibility of conditioning out of a comparison all parameters but the two to be compared was for Rasch the basis of objectivity:

> The term "objectivity" refers to the fact that the result of any comparison of two objects within some specified frame of reference is . . . *independent of everything else within the frame of reference than the two objects which are to be compared and their observed reactions.* (Rasch, 1977; p. 77, italics in original)

The conditional probability in (12) focuses attention on only two observable events. The objective comparison of parameters ξ_j and ξ_k is possible if the distribution of observations over these two events depends only on the difference between ξ_j and ξ_k. This fundamental connection between a pair of parameters and a pair of observable events is the cornerstone of objective measurement. It is the building block out of which all Rasch measurement models are constructed. In this paper, (12) is applied to the particular situation in which observable events form a sequence of ordered alternatives.

3. ORDERED CATEGORIES

Occasionally, responses to test and questionnaire items are recorded in several mutually exclusive response categories intended to represent increasing levels or amounts of the variable being measured. In this case, the observable events (e.g., A, B, C, D) have an intended order $A < B < C < D$. These events may be defined by response alternatives like *strongly disagree, disagree, agree, strongly agree*, or they may be defined as observed stages toward the completion of an item. For convenience, ordered categories of this type will be labeled $0, 1, 2, \ldots, m$, where category 0 indicates the lowest level of response to an item, and category m represents the highest.

The purpose in defining more than two levels of response to an item is to seek more information about a respondent's location β_n on the variable being measured than would be available from the dichotomous scoring of that item. In estimating β_n from observations based on a set of ordered response categories, it is essential that the intended order of these categories be taken into account. The first issue in developing a measurement model for ordered response categories is the question of how this intended order $(0 < 1 < 2, \ldots, < m)$ is to be incorporated into the model. In particular, what are the implications of this intended order for the model probabilities $\pi_{ni0}, \pi_{ni1}, \ldots, \pi_{nim}$ of person n responding in categories $0, 1, \ldots, m$ of item i?

For an item with only two response categories 0 and 1, the implication of category 1 representing "more" of the variable than category 0 is that the probability π_{ni1} of person n responding in category 1 of item i should increase monotonically with β_n. However, for an item with more than two response categories, as β_n increases, the probability of a response in category 1 should decrease as a response in category 2 becomes more likely, and this in turn should decrease as a response in category 3 becomes more likely. For $m > 1$, the implications of the intended order $0 < 1 < \cdots < m$ for the response probabilities $\pi_{ni0}, \pi_{ni1}, \ldots, \pi_{nim}$ are not as simple as for the case $m = 1$.

One approach to specifying the order of a sequence of response

alternatives is through a set of elementary order relations of the form
$A < B$. The smallest number of these elementary relations required to
uniquely define the order of $m + 1$ response categories is m. The
intended order of four categories, for example, is completely captured in
the three elementary order relations $0 < 1$; $1 < 2$; $2 < 3$.

Consider now only response categories 0 and 1. The implication of
the intended order $0 < 1$ is that if a person responds in one of these two
categories, the probability of the response being in category 1 rather than
in category 0 should increase with β_n. In other words, the intended order
of categories 0 and 1 has implications for the *conditional* probability
$\pi_{ni1}/(\pi_{ni0} + \pi_{ni1})$, which should increase with β_n. This observation
suggests a simple application of (12):

$$\frac{\pi_{ni1}}{\pi_{ni0} + \pi_{ni1}} = \frac{\exp(\beta_n - \delta_{i1})}{1 + \exp(\beta_n - \delta_{i1})} \tag{13}$$

In (13) the probability of person n scoring 1 rather than 0 on item i is
governed by only two parameters: the location β_n of person n on the
variable, and a parameter δ_{i1} associated with the transition between
categories 0 and 1 of item i. The form of (13) allows the local comparison
of β_n with δ_{i1} independently of everything else within the frame of
reference (including parameters that might be associated with transitions
between other response categories). Notice that the effect of the
conditioning in (13) is to focus attention on the two observable events
containing information about the relative values of β_n and δ_{i1}. For the
case $m = 1$, (13) is simply Rasch's dichotomous model (10).

Exactly the same reasoning can be applied to each pair of adjacent
response categories $x - 1$ and x. The implication of the intended order
$x - 1 < x$ is that the conditional probability $\pi_{nix}/(\pi_{nix-1} + \pi_{nix})$ should
increase with β_n, suggesting that (13) might be generalized to any pair of
adjacent response categories:

$$\frac{\pi_{nix}}{\pi_{nix-1} + \pi_{nix}} = \frac{\exp(\beta_n - \delta_{ix})}{1 + \exp(\beta_n - \delta_{ix})}, \qquad x = 1, m \tag{14}$$

where δ_{ix} is a parameter associated with the transition (or step) between
response categories $x - 1$ and x. This partial credit model (Masters,
1980, 1982) is simply Rasch's model for objective comparison (12)
applied to the situation in which observable events form a sequence of
ordered alternatives. It focuses on the local comparison of β_n with a
parameter associated with the transition between categories $x - 1$ and x
of item i. The objective comparison of these two parameters is made

possible by conditioning on the two observable events containing information about the relative values of β_n and δ_{ix}.

Local comparisons always take place in a context. In this case, the context includes the full set of response alternatives. In considering this more global context, it is informative to consider the odds of a response in category 1 rather than in category 0:

$$\pi_{ni1}/\pi_{ni0} = \exp(\beta_n - \delta_{i1}) \tag{15}$$

and the odds of a response in category 2 rather than in category 1:

$$\pi_{ni2}/\pi_{ni1} = \exp(\beta_n - \delta_{i2}) \tag{16}$$

It follows directly from (15) and (16) that the odds of a response in category 2 rather than in category 0 are

$$\pi_{ni2}/\pi_{ni0} = \exp(2\beta_n - \delta_{i1} - \delta_{i2}) \tag{17}$$

Similarly,

$$\pi_{ni3}/\pi_{ni0} = \exp(3\beta_n - \delta_{i1} - \delta_{i2} - \delta_{i3}) \tag{18}$$

In fact, odds can be written in this way for any pair or response categories, for example,

$$\pi_{ni4}/\pi_{ni2} = \exp(2\beta_n - \delta_{i3} - \delta_{i4}) \tag{19}$$

showing that, in general, these odds are governed only by β_n and the item parameters corresponding to the transitions between the categories being considered. Under (14) these simple relations apply regardless of the total number of response categories.

When the number of available response alternatives is limited to m, an additional constraint is imposed on the response probabilities, namely,

$$\sum_{k=0}^{m} \pi_{nik} = 1 \tag{20}$$

With this constraint it follows directly that for $m = 3$

$$
\begin{aligned}
\pi_{ni0} &= 1/\Psi \\
\pi_{ni1} &= \exp(\beta_n - \delta_{i1})/\Psi \\
\pi_{ni2} &= \exp(2\beta_n - \delta_{i1} - \delta_{i2})/\Psi \\
\pi_{ni3} &= \exp(3\beta_n - \delta_{i1} - \delta_{i2} - \delta_{i3})/\Psi
\end{aligned}
\tag{21}
$$

where Ψ is the sum of the numerators and ensures that the four response probabilities sum to 1.

These four expressions can be captured in a single general expression if some special notation is introduced. If, for example, we define

$$\sum_{j=0}^{0} (\beta_n - \delta_{ij}) \equiv 0 \quad \text{and} \quad \sum_{j=0}^{k} (\beta_n - \delta_{ij}) \equiv \sum_{j=1}^{k} (\beta_n - \delta_{ij})$$

then

$$\pi_{nix} = \frac{\exp \sum_{j=0}^{x} (\beta_n - \delta_{ij})}{\sum_{k=0}^{m} \exp \sum_{j=0}^{k} (\beta_n - \delta_{ij})}, \qquad x = 0, m \qquad (22)$$

which gives the model probability of person n with location β_n responding in category x ($x = 0, 1, \ldots, m$) of item i.

4. INTERPRETING PARAMETERS

For a given value of m and given values of the item parameters, (22) can be used to obtain model probabilities π_{nix} ($x = 0, 1, \ldots, m$) for a range of values of β_n. Figure 4 shows these probabilities for an item with four response categories ($m = 3$) and item parameters $\delta_{i1} = -1$, $\delta_{i2} = 0$, $\delta_{i3} = 1$. From Figure 4 it is seen that the item parameters $\delta_{i1}, \delta_{i2}, \ldots, \delta_{im}$ in (22) have a simple interpretation: they correspond to the intersections of adjacent probability curves.

This simple interpretation of the item parameters in the partial credit

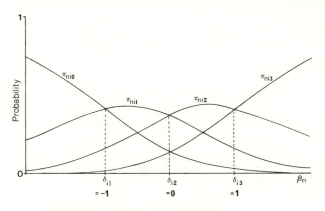

FIGURE 4. Model probability curves for $m = 3$.

model is very convenient when it comes to studying the operational definition of variables. In Figure 4, the three item parameters mark out regions of most probable response to this item. Once a person's location β_n has been estimated, this estimate can be referred to the item estimates to read off that person's most probable response to this item.

The observation that all person and item parameters in (22) are locations on the variable being measured is another feature that this model shares with the dichotomous Rasch model. This feature sets these models apart from other models that incorporate discrimination or dispersion parameters that are not locations, but which serve only to qualify location parameters.

An important distinction can be drawn between the item parameters (δ_{ij}) in (22) and another set of parameters sometimes used in models for ordered response categories. Consider, for example, Samejima's (1969) graded response model for an item with response categories $0, 1, \ldots, m$:

$$\sum_{j=k}^{m} \pi_{nij} = \frac{\exp[\alpha_i(\beta_n - \gamma_{ik})]}{1 + \exp[\alpha_i(\beta_n - \gamma_{ik})]}, \qquad k = 1, m \qquad (23)$$

The item parameters (γ_{ik}) in this model are usually referred to as "category boundaries." However, rather than being based on local comparisons at transitions between adjacent response categories, these parameters are based on a global comparison in which the item parameter γ_{ik} governs the probability of a response *in or above* category k. Category boundaries defined in this way do not mark out regions of most probable response and do not permit the separation of person and item parameters necessary for objective comparison (Molenaar, 1983).

The difference between the item parameters of (22) and Samejima's global "category boundaries" can be seen from Table 1. Whereas the δ_{ij}'s are defined locally, category boundaries are defined in terms of

TABLE 1

A Comparison of Item Parameters in the Partial Credit and Graded Response Models ($m = 3$)

Partial credit model		Graded response model	
Parameter	Location	Parameter	Location
δ_{i1}	$\pi_{ni0} = \pi_{ni1}$	γ_{i1}	$\pi_{ni0} = \pi_{ni1} + \pi_{ni2} + \pi_{ni3}$
δ_{i2}	$\pi_{ni1} = \pi_{ni2}$	γ_{i2}	$\pi_{ni0} + \pi_{ni1} = \pi_{ni2} + \pi_{ni3}$
δ_{i3}	$\pi_{ni2} = \pi_{ni3}$	γ_{i3}	$\pi_{ni0} + \pi_{ni1} + \pi_{ni2} = \pi_{ni3}$

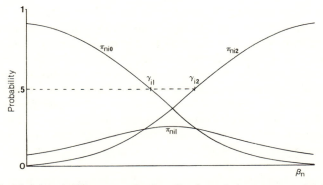

FIGURE 5. Model probability curves for Samejima's graded response model ($m = 2$).

cumulative probabilities. A consequence of this cumulative definition is that category boundaries are, by definition, ordered $\gamma_{i1} < \gamma_{i2} < \cdots < \gamma_{im}$. This can be seen from Figure 5 where the response probabilities for Samejima's model are plotted for an item with three ordered categories. The item parameters γ_{i1} and γ_{i2} correspond to locations at which $\pi_{ni0} = \pi_{ni1} + \pi_{ni2} = .5$ and $\pi_{ni0} + \pi_{ni1} = \pi_{ni2} = .5$. As the difference $\gamma_{i2} - \gamma_{i1}$ decreases, so does every person's probability π_{ni1} of responding in category 1. Provided that this probability is nonnegative, $\gamma_{i1} < \gamma_{i2}$ by definition.

Figure 5 can be contrasted with a similar probability plot for the partial credit model (Figure 6). Notice that δ_{i1} and δ_{i2} correspond to the intersections of the probability curves for response categories 0 and 1, and 1 and 2. However, δ_{i1} and δ_{i2} are no longer ordered $\delta_{i1} < \delta_{i2}$. This brings out an important distinction between the item parameters in the

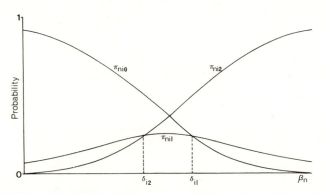

FIGURE 6. Model probability curves for partial credit model ($m = 2$).

partial credit and graded response models. In the partial credit model, the order of the response categories is defined in terms of the elementary order relations $0 < 1, 1 < 2, \ldots, m - 1 < m$, and the item parameters $\delta_{i1}, \delta_{i2}, \ldots, \delta_{im}$ govern the transitions between adjacent response categories. Order is not incorporated through the values of these locally defined parameters, which are in fact free to take any values at all.

This feature of the partial credit model can be understood by imagining an item that is completed in two parts, the second half of the item being attempted only after the first half has been successfully completed. Such an item defines three performance levels: failure on the first half ($x = 0$), success on the first half but failure on the second ($x = 1$), and success on the first half followed by success on the second ($x = 2$). The fact that the two parts of this item must be completed in order has no implication for the relative difficulties of those two halves. The second half may in fact be easier than the first. If this is the case, then the probability of a person completing the first half but being unable to complete the second (and so scoring 1 on the item) will be low. This is the situation depicted in Figure 6, where a score of 1 is never a very probable result on this item.

It should be noted, however, that for a two-part item, the two item parameters δ_{i1} and δ_{i2} in the partial credit model *cannot* be interpreted simply as the "difficulties" of the two independent halves of that item. This is because of their conditional definition. They are more appropriately interpreted as a set of parameters associated with item i, none of which can be interpreted meaningfully outside the context of the entire item in which it occurs.

It should also be noted that while the observation $\hat{\delta}_{i2} < \hat{\delta}_{i1}$ does not indicate a violation of the partial credit model, as $\hat{\delta}_{i1} - \hat{\delta}_{i2}$ becomes more negative, a response in category 1 becomes less likely, meaning that the scoring of item i becomes increasingly dichotomous as students who do not score 0 on this item become increasingly likely to score 2.

5. RELATED MODELS

A particularly convenient feature of the partial credit model (22) is that it provides a framework for illustrating the relationships among a number of other measurement models for ordered response categories. With (22) as a point of reference, these models can be interpreted as versions of this general model in which constraints are imposed on the values the item parameters can take. These constraints usually are suggested by the way in which the response categories are defined.

5.1. Rating Scales

Ordered response categories are sometimes defined by a set of alternatives like *poor/fair/good/excellent*, and *never/sometimes/often/always*, which are intended for use with all items on a test or questionnaire. When response categories are defined in this way, it might be hypothesized that the functioning of these alternatives should not vary significantly from item to item. This expectation can be incorporated into (22) by resolving the set of parameters (δ_{ij}) for each item into a single item parameter and a set of parameters for the response alternatives:

$$\delta_{ij} = \delta_i + \tau_j, \qquad j = 1, m \qquad (24)$$

With this rating scale constraint, (22) becomes the rating scale model (Andrich, 1978; Masters, 1980) in which a single location δ_i is estimated for each item, and m parameters τ_j $(j = 1, m)$ are estimated for the $m + 1$ response alternatives. (It is usual to define δ_i as the mean of the δ_{ij}'s so that $\sum \tau_j = 0$.)

The role of the parameters in the rating scale model is illustrated in Figure 7. Under constraint (24), the pattern of δ_{ij}'s (defined by the parameters $\tau_1, \tau_2, \ldots, \tau_m$) is held constant across items and is assumed to depend only on the provided response alternatives. The only modeled difference between items is in their location δ_i on the variable. When data approximate the rating scale model it offers the advantage of involving fewer parameters than (22). The cost of this parsimony may be the loss of information about subtle but psychologically interesting interactions between items and response alternatives. Applications of this model to a wide variety of rating scales have been described by Andrich (1978, 1979), Duncan (1984, 1985), Masters (1980), and Wright and Masters (1982).

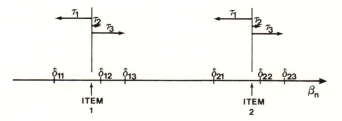

FIGURE 7. The rating scale constraint.

5.2. BINOMIAL TRIALS

A rating scale is an example of one particular way of defining ordered response categories. A second way in which ordered categories might be defined is as counts of successes in m attempts at an item. In most educational tests, only one attempt is allowed at each item and, if more than one attempt were to be allowed, then we would expect a person's probability of success to be different on different attempts as alternative strategies were tried out and proved fruitless. Nevertheless, the definition of response categories as counts of successes on m independent attempts at an item provides a useful point of reference in the study of other response formats. It is also considered here because this model has been the subject of some discussion in the literature (Andrich, 1978; Rasch, 1972).

The binomial trials model is developed from Rasch's model for the probability π_{ni} of person n succeeding on a single attempt at dichotomously scored item i. If person n is allowed m attempts at item i, and these m attempts are assumed independent, then person n's probability of succeeding on x of m attempts is

$$\pi_{nix} = \binom{m}{x} \pi_{ni}^{x} (1 - \pi_{ni})^{m-x} \qquad (25)$$

If Rasch's dichotomous model for the probability π_{ni} of person n succeeding on a single attempt at item i is substituted from (10) into (25) then

$$\pi_{nix} = \frac{\exp \sum_{j=0}^{x} (\beta_n - (\delta_i + c_j))}{\sum_{k=0}^{m} \exp \sum_{j=0}^{k} [\beta_n - (\delta_i + c_j)]}, \qquad x = 0, m \qquad (26)$$

where

$$c_j = \ln[j/(m - j + 1)], \qquad \sum_{j=0}^{0} [\beta_n - (\delta_i + c_j)] \equiv 0$$

and

$$\sum_{j=0}^{k} [\beta_n - (\delta_i + c_j)] \equiv \sum_{j=1}^{k} [\beta_n - (\delta_i + c_j)]$$

A more complete derivation of (26) is provided by Masters and Wright (1984).

From (26) it is seen that the binomial trials model can also be thought of as a version of (22) in which a particular constraint is imposed on the item parameters (δ_{ij}). In this case, the constraint is a consequence of defining ordered response categories as counts of successes on m independent attempts at an item with difficulty δ_i:

$$\delta_{ij} = \delta_i + \ln[j/(m - j + 1)] \tag{27}$$

Under this binomial trials constraint, the pattern of δ_{ij}'s is not estimated as in the rating scale model, but is fixed across items and follows directly from the assumption that the m attempts at item i are independent Bernoulli trials. For an item with four ordered categories, for example, the distance between δ_{i1} and δ_i is defined as $c_1 = \ln[1/(3 - 1 + 1)] = 1.1$ logits.

5.3. POISSON COUNTS

Under the binomial response format just considered, the number of successes on an item has some finite upper limit m. However, in some testing contexts, there is no obvious limit to the number of levels of performance on an item that might be observed. In assessing a child's reading fluency, for example, a count may be made of the number of errors the child makes on a particular passage of text. These counts define ordered levels of performance, but there is no particular limit to the number of errors that might be observed. In this context, it is convenient to introduce the Poisson model as the limiting case of the binomial.

To develop the Poisson model we continue to count successes rather than errors and consider the situation in which item i presents very many opportunities for success, but the probability of success on any particular opportunity is small. Under the binomial model, person n's expected number of successes on item i is $\lambda_{ni} = m\pi_{ni}$, where m is the number of opportunities for success and π_{ni} is person n's probability of success on any given opportunity. If, as m becomes large and π_{ni} becomes small, the expected number of successes λ_{ni} remains constant, then in the limit, the binomial model becomes the Poisson

$$\pi_{nix} = \frac{\lambda_{ni}^x}{x! \exp(\lambda_{ni})} \tag{28}$$

in which the person parameter β_n and item parameter δ_i have been replaced by a single parameter λ_{ni} representing person n's expected number of successes on item i. It follows directly from (28) that the conditional probability of person n scoring 1 rather then 0 on item i is

$$\frac{\pi_{ni1}}{\pi_{ni0} + \pi_{ni1}} = \frac{\lambda_{ni}}{1 + \lambda_{ni}} \tag{29}$$

The form of (29) suggests a simple interpretation of λ_{ni} which would render (29) equivalent to (13) and express person n's expected number of successes on item i in terms of a parameter for person n and a parameter for item i. Substituting $\lambda_{ni} = \exp(\beta_n - \delta_{i1})$ into (28) yields

$$\pi_{nix} = \frac{\exp[x(\beta_n - \delta_{i1})]}{x! \exp[\exp(\beta_n - \delta_{i1})]} \tag{30}$$

where δ_{i1} governs the probability of a score of 1 rather than 0 on item i. This is the Poisson model developed by Rasch (1960). It can be expressed in the partial credit form by substituting the definitions

$$x! = \exp\left(\sum_{j=1}^{x} \ln j\right) \quad \text{and} \quad \exp(\lambda) = \sum_{k=0}^{\infty} \lambda^k / k!$$

into (30) to obtain

$$\pi_{nix} = \frac{\exp \sum_{j=0}^{x} [\beta_n - (\delta_{i1} + \ln j)]}{\sum_{k=0}^{\infty} \exp \sum_{j=0}^{k} [\beta_n - (\delta_{i1} + \ln j)]}, \qquad x = 0, \infty \tag{31}$$

which is equivalent to (22) with $m = \infty$ and

$$\delta_{ij} = \delta_{i1} + \ln j \tag{32}$$

(Poisson counts constraint).

Figure 8 shows the Poisson model probability curves for the first seven response categories of item i. This pattern continues indefinitely to the right, with the probability curve for each response category being slightly lower than the preceding curve on its left. The same pattern of probability curves applies to each item, with only the location δ_{i1} of this pattern varying from item to item.

5.4. OTHER CONSTRAINTS

In the rating scale, binomial trials, and Poisson counts models constraints on the δ_{ij}'s follow from the special ways in which ordered

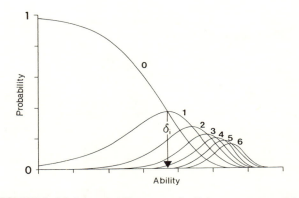

FIGURE 8. Model probability curves for Poisson counts model.

response categories are defined by these formats. But there is no limit to the number of ways in which the δ_{ij}'s in (22) might be constrained and reparametrized to produce other special cases of this general model. Other models have been proposed in which the δ_{ij}'s are constrained to be uniformly spaced: either with a fixed uniform spacing (Andrich, 1978), a uniform spacing that is estimated once for all items (Masters, 1980), or separately for each item (Andrich, 1982), or in which the spacing of the δ_{ij}'s is constrained to increase or to decrease in a particular direction. In general, the introduction of constraints on the item parameters in (22) is probably justified only if it has some psychological basis or arises from the way in which ordered response categories are defined.

6. REFERENCES

Andrich, D. (1978). A rating formulation for ordered response categories. *Psychometrika*, *43*, 561–573.

Andrich, D. (1979). A model for contingency tables having an ordered response classification. *Biometrics*, *35*, 403–415.

Andrich, D. (1982). An extension of the Rasch model for ratings providing both location and dispersion parameters. *Psychometrika*, *47*, 105–113.

Duncan, O. D. (1984). Rasch measurement in survey research: Further examples and discussion. in C. F. Turner & E. Martin (Eds.), *Surveying subjective phenomena*. New York: Russell Sage Foundation.

Duncan, O. D. (1985). Some models of response uncertainty for panel analysis. *Social Science Research*, *14*, 126–141.

Masters, G. N. (1980). A Rasch model for rating scales. *Dissertation Abstracts International*, *41*, 215A–216A.

Masters, G. N. (1982). A Rasch model for partial credit scoring. *Psychometrika*, *47*, 149–174.

Masters, G. N., & Wright, B. D. (1984). The essential process in a family of measurement models. *Psychometrika, 49,* 529–544.

Molenaar, I. (1983). Item steps. *Heymans Bulletins Psychologische* (Report No. HB-83-630-EX). Gronigen, Netherlands: Instituten R.U.

Rasch, G. (1960). *Probabilistic models for some intelligence and attainment tests.* Copenhagen: Denmarks Paedagogiske Institut.

Rasch, G. (1972). Objectivity in the social sciences: A methodological problem [in Danish], *Nationalekonomisk Tiddskrift, 110,* 161–196.

Rasch, G. (1977). On specific objectivity: An attempt at formalizing the request for generality and validity of scientific statements. *Danish Yearbook of Philosophy, 14,* 58–94.

Samejima, F. (1969). Estimation of latent ability using a response pattern of graded scores. *Psychometrika* (Monograph Supplement No. 17).

Wright, B. D., & Masters, G. N. (1982). *Rating scale analysis.* Chicago: MESA Press.

CHAPTER TWO

Testing a Latent Trait Model

ARNOLD L. VAN DEN WOLLENBERG

1. INTRODUCTION

Within the domain of latent trait models the Rasch model (Rasch, 1960) takes a prominent place. This fact can be accounted for by the special characteristics of the model deriving from specific objectivity. As a consequence, this one-parameter logistic model, as it is also called, has been studied extensively for the past decade. Also with respect to model testing the Rasch model has been studied more thoroughly than other latent trait models. For this reason we will concentrate in the present paper on testing the Rasch model. Most of the points made with respect to the Rasch model apply to other latent trait models as well.

We will offer a fairly extensive description of test statistics and procedures. The statistics such as those of Andersen (1973b), Wright and Panchapakesan (1969), Fischer and Scheiblechner (1970), Martin-Lof (1973), and van den Wollenberg (1979, 1982b) can be reduced to basically two types. All statistics test for sample independence, with the exception of the Q_2 statistic of van den Wollenberg, which tests for unidimensionality and local independence. The statistics testing for sample independence will be referred to as statistics of the Q_1 type. They will be discussed in Section 3.

Testing the unidimensionality axiom has been neglected until recently. In Section 4 methods to investigate the dimensionality axiom are presented.

ARNOLD L. VAN DEN WOLLENBERG • Department of Mathematical Psychology, University of Nijmegen, 6500HE Nijmegen, The Netherlands. Preparation of this manuscript was supported by Grant No. 00-56-243 from the Dutch Science Foundation ZWO.

Descriptive goodness-of-fit measures have been given little attention up to now, yet they can be a valuable tool in the evaluation of latent trait models. As a rule one is not (or if so, one should not be) interested in the question whether a model fits the data, but rather whether the model gives a good approximation of the data structure. With respect to this question, descriptive goodness-of-fit measures can in addition contribute to statistical model testing. They are presented in Section 5.

2. THE MODEL

The Rasch model can be formulated in terms of the item characteristic function, which is one-parameter logistic:

$$\pi_{vi} = f_i(\xi_v) = \frac{\exp(\xi_v - \sigma_i)}{1 + \exp(\xi_v - \sigma_i)} \tag{1}$$

where ξ_v is the subject parameter and σ_i is the item parameter of the Rasch model. For estimation and testing purposes, however, a reparametrization of (1) is used in which $\theta_v = \exp(\xi_v)$ and $\varepsilon_i = \exp(-\sigma_i)$:

$$\pi_{vi} = \frac{\theta_v \varepsilon_i}{1 + \theta_v \varepsilon_i} \tag{2}$$

From the fact that the item is characterized by only one parameter, the model is said to be one-parameter logistic.

It can be shown that the above formulations are (up to a monotone scale transformation) equivalent to the following set of axioms (e.g., Fischer, 1974):

1. *Monotonicity.* The probability of a positive response increases strictly as a function of the latent trait.
2. *Sufficiency.* The number of positive responses of a subject is a sufficient statistic for the subject parameters; the number of positive responses to a given item is a sufficient statistic for the item parameter.
3. *Unidimensionality.* Items forming a Rasch homogeneous scale appeal to just one underlying dimension.
4. *Local Stochastic Independence.* For fixed latent trait all items are independent.

These axioms are necessary and sufficient for the Rasch model and thus can be used interchangeably with (1) or (2).

The Rasch model possesses several desirable properties. We will concentrate on just one property that is important in the course of testing the model; this property is called sample independence. Sample independence implies that estimation of the item parameters is not affected by the selection of subjects: for every sample from the Rasch homogeneous universe, the item parameters are the same; so over samples the estimated item parameters should be the same within chance limits. It is this property that, until recently, has been used as almost the sole means of testing the Rasch model.

On the level of estimation, sample independence is guaranteed by the conditional maximum likelihood (CML) method. In CML estimation use is made of the basic symmetric functions and their partial derivatives. The basic symmetric functions are defined as the following functions of the item parameters:

$$\gamma_0 = 1$$
$$\gamma_1 = \varepsilon_1 + \varepsilon_2 + \cdots + \varepsilon_k$$
$$\gamma_2 = \varepsilon_1\varepsilon_2 + \varepsilon_1\varepsilon_3 + \cdots + \varepsilon_{k-1}\varepsilon_k \qquad (3)$$
$$\vdots$$
$$\gamma_k = \varepsilon_1\varepsilon_2\varepsilon_3 \cdots \varepsilon_k$$

where k is the number of items. γ_r is said to be the basic symmetric function of order r. The partial derivative of the basic symmetric function with respect to the item parameter ε_i, then, is defined as

$$\partial\gamma_r/\varepsilon_i = \gamma_{r-1}^{(i)} \qquad (4)$$

The partial derivative of the basic symmetric function of order r with respect to item parameter ε_i is the basic symmetric function of order $r - 1$ in all item parameters to the exclusion of ε_i (see Andersen, 1973a; or Fischer, 1974).

By means of the basic symmetric functions the probability of a positive response on an item, given score r, can be estimated independently of the subject parameter:

$$p(+|i, v, r) = \hat{\pi}_{vi} = \frac{\hat{\varepsilon}_i \times \hat{\gamma}_{r-1}^{(i)}}{\hat{\gamma}_r} \qquad (5)$$

Now the likelihood of the total sample can be obtained independently of the subject parameters:

$$L = p\{(n_{.i}) \mid (n_{v.}), (\hat{\theta}), (\hat{\varepsilon})\} = \begin{bmatrix} (n_{v.}) \\ (n_{.i}) \end{bmatrix} \frac{\prod \hat{\varepsilon}_i^{n_{.i}}}{\prod \hat{\gamma}_{n_{v.}}(\hat{\varepsilon})} \tag{6}$$

in which $(n_{.i})$ and $(n_{v.})$ are vectors of marginal frequencies of the data matrix and (ε) and (θ) are the item and subject parameter vectors, respectively. $[\binom{n_v}{n_i}]$ is a combinatorial quantity giving the total number of possible data matrices with given marginals. For a more detailed presentation of the conditional likelihood method see Andersen (1973a) or Fischer (1974). In the present context we will only make use of (6) in the presentation of the conditional likelihood ratio test.

3. TESTS FOR SAMPLE INDEPENDENCE

Sample independence implies that estimated item parameters should, within chance limits, be the same in each sample from a Rasch homogeneous universe. So when a given sample is split up into subsamples the same equality should hold over the subsamples. All tests for sample independence are, more or less directly, tests for the equality of item parameters over subsamples.

The most common split-up of a sample is that according to raw score; that is, all subjects with the same raw score form a subsample. In this way $k - 1$ subsamples can be obtained (subjects passing all or none of the items are excluded from the analysis, as they do not bear information with respect to the item parameters). In principle any raw score partitioning can be used, grouping score groups together; for instance, the split-up of the sample in a high- and a low-scoring group is very popular. When a raw score partitioning is used, the parallelism of the item characteristic curves is in effect being tested (Gustafsson, 1980; van den Wollenberg, 1979).

Besides raw score partitioning, any other partitioning is allowed; for instance according to demographic variables such as sex, age, and the like or according to a test item as in the splitter item technique below.

3.1. THE CONDITIONAL MAXIMUM LIKELIHOOD RATIO TEST

In conjunction with the CML estimation method Andersen (1973b) offered a test statistic, the conditional likelihood ratio test, which follows standard likelihood ratio principles. Andersen (1973b) introduces the statistic using the complete raw score partitioning. The total likelihood is obtained according to (6); analogously the likelihood of each subsample can be obtained. It can be shown (Andersen, 1973b) that, under the model, the total likelihood is equal to the product of likelihoods of the subsamples. So the likelihood ratio

$$\lambda = \frac{L}{\Pi L_r} \qquad (r = 1, \ldots, k - 1) \tag{7}$$

is equal to 1. When the likelihoods are estimated, this equality only holds approximately. Andersen (1973b) showed that in this case the quantity $Z = -2 \ln \lambda$ is distributed as χ^2.

The likelihood ratio departs from 1 insofar as the item parameters differ over groups, in other words Z tests for sample independence.

3.2. THE MARTIN-LOF TEST

The Martin-Lof (1973) statistic is a Pearsonian chi-squared statistic. In this statistic the observed frequencies of correct answers are compared with the expected frequencies obtained by means of (5):

$$T = \sum_{r=1}^{k-1} (d_r)'(V_r)^{-1}(d_r) \tag{8}$$

with $(V_r)^{-1}$ the variance–covariance matrix of the conditional expectations $E(n_{ri} \mid n_r)$ and vector (d_r):

$$(d_r) = (n_r) - [E(n_r)], \qquad E(n_{ir}) = \frac{n_r \hat{\varepsilon}_i \hat{\gamma}_{r-1}^{(i)}}{\hat{\gamma}_r} \tag{9}$$

It is apparent that the Martin-Lof statistic tests for equality of item parameters directly, whereas Andersen's Z statistic is an indirect test of the same property. The statistics are sensitive to the same model deviations and are asymptotically equivalent.

3.3. THE VAN DEN WOLLENBERG STATISTIC Q_1

The Q_1 statistic (van den Wollenberg, 1979, 1982b) can be looked upon as a simplification of the Martin-Lof statistic. If it is assumed that the off-diagonal values of the variance–covariance matrix $(V_r)^{-1}$ are equal, then Q_1 results:

$$Q_1 = \frac{k-1}{k} \sum_i q_i \tag{10}$$

with

$$q_i = \sum_r^{k-1} \left\{ \frac{[n_{ri} - E(n_{ri})]^2}{E(n_{ri})} + \frac{[n_{ri} - E(n_{ri})]^2}{n_r - E(n_{ri})} \right\} \qquad (11)$$

The statistics Q_1 and T are very much in agreement; van den Wollenberg (1979, 1982a, 1982b) reports correlations of .99 and even higher, while the mean values and variances are also practically equal. For all practical purposes these statistics can be considered equivalent.

3.4. THE FISCHER–SCHEIBLECHNER STATISTIC S

The statistics mentioned above are applicable to a split-up of the sample according to raw score; in addition Z and Q_1 can be applied to any split-up of the data set into subsamples. The Fischer–Scheiblechner (1970) procedure presupposes a split-up of the sample into two sub-samples. In each of the subsamples the item parameters are estimated and the estimates are compared by means of a z test:

$$S_i = \frac{\hat{\sigma}_i^{(1)} - \hat{\sigma}_i^{(2)}}{[s_{\hat{\sigma}_i^{(1)}}^2 + s_{\hat{\sigma}_i^{(2)}}^2]^{1/2}} \qquad (12)$$

where $\hat{\sigma}_i^{(1)}$ and $\hat{\sigma}_i^{(2)}$ stand for the estimated parameter in the first and second subsample, respectively. S_i is an approximate standard normal variate, so S_i^2 is distributed as an approximate chi-squared. Now the overall statistic S is formed by adding individual S_i^2:

$$S = \sum_i S_i^2 \qquad (i = 1, \ldots, k) \qquad (13)$$

Now S is a sum of chi-squared terms and, according to Fischer and Scheiblechner, S is distributed as chi-squared with $k - 1$ degrees of freedom. However, as van den Wollenberg (1979) points out, this reasoning is inconsistent. The expectation of S as defined by Fischer and Scheiblechner is equal to k, k terms all with expectation 1. So expectation and degrees of freedom are not in accordance. Furthermore, simple addition of the individual S_i^2 is allowed when the chi-squared terms are independent. This is not the case here, as the items are normed and thereby negatively correlated. The covariance between the estimated item parameters is ignored. S cannot be distributed as chi-squared with $k - 1$ degrees of freedom nor as chi-squared with k degrees of freedom.

3.5. THE WRIGHT–PANCHAPAKESAN STATISTIC Y

The statistical model test of Wright and Panchapakesan (1969) entails a comparison of expected and observed freuencies for every item in every level group, but just like the Z and Q_1 statistics other split-ups are feasible. A standard deviate is defined, which is

$$y_{ri} = \frac{n_{ri} - E(n_{ri})}{[V(n_{ri})]^{1/2}} \tag{14}$$

Here n_{ri} is the number of subjects with raw test score r (level group r) that respond positively to item i, and $E(n_{ri})$ is the corresponding expectation, whereas $V(n_{ri})$ is the variance. Expectation and variance are defined as

$$y_{ri} = n_r \times \pi_{ri}, \qquad V(n_{ri}) = n_r \times \pi_{ri} \times (1 - \pi_{ri}) \tag{15}$$

with π_{ri} the probability of a positive response to item i of a subject with test score r. When the parameters of the model have been estimated, the probability, π_{ri}, is estimated as

$$\hat{\pi}_{ri} = \frac{\exp(\hat{\xi}_r - \hat{\sigma}_i)}{1 + \exp(\hat{\xi}_r - \hat{\sigma}_i)} \tag{16}$$

where $\hat{\xi}_r$ is the estimated parameter for a subject with test score r.

Wright and Panchapakesan (1969) state that the statistic y_{ri} will have an approximate unit normal distribution, provided the model holds and n_r is large enough. So the squares of the approximate unit normal deviates will be approximate χ^2 distributed variates. An overall statistic is obtained as

$$Y = \sum_r^m \sum_i^k y_{ri}^2 \tag{17}$$

Wright and Panchapakesan state that this statistic is approximately χ^2 distributed with $(k - 1)(m - 1)$ degrees of freedom, with k the number of items and m the number of level groups with $n_r \neq 0$.

However, when each statistic y_{ri} is a unit normal deviate, each y_{ri}^2 is a chi-squared distributed variate with one degree of freedom. Then the overall statistic Y is a sum of $k \times m$ terms each with expectation 1. In other words

$$E(Y) = \sum_r^m \sum_i^k E(y_{ri}^2) = \sum_r^m \sum_i^k 1 = k \times m \tag{18}$$

This expectation of Y is inconsistent with the statement that Y is distributed as χ^2 with $(m - 1)(k - 1)$ degrees of freedom. van den Wollenberg (1980) showed that the individual y_{ri} are not unit normal deviates: because of estimation, degrees of freedom are lost and, furthermore, because of norming, the parameter estimates are not independent, just as in the Fischer–Scheiblechner statistic.

A more serious point bears upon the equation for estimating the probability π_{ri}. In the unconditional equation (16) only the parameter of item i is incorporated, whereas the conditional equation (5) takes all item parameters into account by means of the symmetric functions and their derivatives. This implies that equation (16) does not take the difficulties of the other items in the set into account as it should. As a consequence, the estimated probabilities can be biased, as is illustrated below.

Consider the two sets of item parameters shown in Table 1 (we use true item parameters instead of estimates to stress the fact that the bias is dependent upon a wrong probability formula and has nothing to do with the method of estimation):

TABLE 1
Two Sets of Parameters

Item	σ_i	$\varepsilon_i = \exp(-\sigma_i)$	Item	σ_i	$\varepsilon_i = \exp(-\sigma_i)$
1	0.0	1.0	1	−2.0	7.3891
2	0.0	1.0	2	−2.0	7.3891
3	0.0	1.0	3	0.0	1.0000
4	0.0	1.0	4	4.0	0.0183

The first set consists of items that are equally difficult, whereas the second set of items shows considerable variation in item difficulty. When a given subject has raw score 2, the subject parameter estimate in the first set of items will be $\hat{\xi}_v = 0.0$. When the algorithm described by Wright and Panchapakesan (1969, p. 41) is used, application of the equations (16) and (5) gives in both instances 0.5 as the estimated probability for any item; equations (16) and (5) are in accordance, as they should be.

For the second item set the results are not as good. The estimated subject parameter now becomes $\hat{\xi}_v = -.5433$, and the estimated probability for item 3 is .367, when the Wright–Panchapakesan formula is applied. Use of formula (6) give a quite different estimate of .212. If the number of subjects with raw score 2 were 100, the deviation between the two estimated probabilities would give rise to a chi squared of 14.38 for this item-level combination.

TABLE 2

The Behavior of the Statistics Y and Q_1 for Equal and Unequal Item Parameters; 4000 Subjects, 100 Replications, Data Conforming to the Rasch Model

	Number of items				
	4	5	6	7	8
Equal item parameters					
df	6	12	20	30	42
Y	6.04	12.67	19.64	30.98	43.33
Q_1	5.82	12.50	19.47	30.85	43.25
$2 \times df$	12	24	40	60	84
$s(Y)$	9.91	21.60	34.63	58.38	94.90
$s(Q_1)$	9.90	21.77	34.56	58.54	95.08
r	.9991	.9998	.9998	.9999	1.000
Unequal item parameters					
Parameters	$(-2, -1)$ $(-1, 2)$	$(-2, -1)$ $(0, 1, 2)$	$(-2, -1, -1)$ $(1, 1, 2)$	$(-2, -1, -1)$ $(0, 1, 1, 2)$	$(-2, -1, 1, 2)$ $(-2, -1, 1, 2)$
df	6	12	20	30	42
Y	426.03	293.89	211.15	192.88	245.73
Q_1	5.63	11.94	20.83	28.55	41.46
$2 \times df$	12	24	40	60	84
$s(Y)$	373.87	350.06	181.03	145.56	184.22
$s(Q_1)$	11.14	21.59	50.25	68.61	83.90
r	.0840	.3510	.3142	.6100	.6459

When the flaws in the Wright–Panchapakesan statistic are corrected, the Q_1 statistic results. In Table 2 the possible effect of the wrong probability formula is illustrated; the other two flaws mentioned above have been corrected, so the effect is solely due to the use of (16) instead of the correct formula (5). It can be observed that for the case of equal item parameters, the statistics are practically equivalent (theoretically they are; the differences are algorithmically based). When the item parameters differ, Y shows considerable deviations from the chi-squared distribution; these deviations decrease when the number of items increases.

3.6. COMPARISON OF TESTS FOR SAMPLE INDEPENDENCE

Of the five test statistics described above, both the Fischer and Scheiblechner and the Wright and Panchapakesan statistics suffer from rather serious theoretical flaws. As other statistics can do the same job, there seems to be no reason to use these statistics any longer.

Of the remaining three statistics, the likelihood ratio test seems to be preferable from a theoretical point of view. As estimation proceeds along the lines of maximum likelihood, a likelihood statistic is the most appropriate. Both T and Q_1 are of the Pearson chi-squared type. Q_1 must be looked upon as an approximation of T based on a simplifying assumption. So from a theoretical point of view, Z is superior to T, which in turn is superior to Q_1.

van den Wollenberg (1979, 1982b,c) showed in numerous simulations with different parameter configurations that the correspondence between these three statistics is very high. He found correlations between Z on the one hand and Q_1 and T on the other, which well exceeded .95. The correlations between Q_1 and T were even beyond .99. The statistics show values that can be taken as equivalent for all practical purposes.

When the three statistics are compared with respect to their practicality, the order is clearly reversed.

1. Applicability. For the likelihood ratio test parameters have to be estimated in each of the subsamples of the partitioning. This will prove impossible when one of the items is passed or failed by all subjects. This may be quite likely, when in a small sample a partitioning according to raw score is used.

In the course of obtaining T an inverse has to be obtained in each level group, which also may be troublesome. For the Q_1 statistic only the overall parameter estimates have to be obtained and no intricate calculations have to be performed. So when the overall parameters can be estimated, Q_1 can be obtained for any partitioning of the sample.

2. Computational Efficiency. Because only the overall parameters are needed and no inverses have to be obtained, Q_1 can be obtained with less computation time than Z and T.

3. Diagnostic Power. In Q_1 a chi-squared contribution is obtained for each item-level combination. This is extremely important in evaluating lack of fit. This possibility is completely absent for Z, and for T it is not practical, as here chi-squared contributions can become negative.

It should be clear that Q_1 is by far the most practical of these three statistics. Because the sampling behavior of the three statistics is such that no differences of any importance can be pointed out, it seems straightforward to recommend the Q_1 statistic. This seems the more justified as Q_1, together with its counterpart Q_2, constitutes a complete test of the Rasch model. The other statistics lack such a counterpart.

3.7. OTHER PROCEDURES FOR TESTING SAMPLE INDEPENDENCE

Above we discussed the most commonly used statistics for testing sample independence in the Rasch model. Other developments must be mentioned for the sake of completeness:

Rasch (1960) introduced a parameter-free test, which can be looked upon as a generalization of the Fisher exact test.

Other types of model tests concentrate on a single item, making use of the binomial distribution; relevant references are Allerup and Sorber (1977), Gustafsson (1977), and Molenaar (1983).

Glas (1981) introduced a Q_1 statistic for tests with missing data. Kelderman (1984) showed that the Q_1 test (just like the Q_2 test to follow) can also be formulated in terms of a log-linear model.

4. TEST FOR DIMENSIONALITY AND LOCAL INDEPENDENCE

All the above tests inspect item equality over subsamples and thereby are especially sensitive to violations of the axioms of sufficiency and monotonicity. Gustafsson (1980), Stelzl (1979), and van den Wollenberg (1979, 1982b) showed that under certain circumstances these tests may be insensitive to violation of the axioms of local independence and unidimensionality. van den Wollenberg gave a sufficient set of conditions for a multidimensional data structure to behave perfectly Rasch homogeneously in the sense of the tests of sample independence.

Up to very recently all Rasch model testing has been performed by means of tests for sample independence, and even nowadays dimensionality testing is often ignored. It seems a rather distressing fact that the important axiom of unidimensionality is not as a rule investigated. Therefore, many applications of the Rasch model may well be invalid.

Until recently the only means of testing dimensionality in the Rasch model was a procedure proposed by Martin-Lof (1973). However, this procedure presupposes an *a priori* hypothesis as to the dimensional structure of the test, which as a rule is not available. This method has therefore hardly been applied. van den Wollenberg (1979, 1982a, 1982b) introduced both a statistic and a technique to investigate the dimensionality and independence assumptions.

4.1. THE MARTIN-LOF PROCEDURE

When a set of items is unidimensional Rasch homogeneous and when it it split up into two subtests consisting of k_1 and k_2 items, respectively, it can be shown tht the likelihood of the total test is a function (which we will not specify explicitly) of the likelihoods of the subtests:

$$L = f(L_1, L_2) \tag{19}$$

This function is distributed as chi squared with $k_1 k_2 - 1$ degrees of freedom.

When the two subtests measure separate traits, the overall likelihood exceeds the function of the separate likelihoods, which will be reflected in a likelihood ratio test.

It is obvious that some *a priori* knowledge should be available on the composition of the subtests. As a rule this knowledge will not be available.

4.2. The Q_2 Statistic

When a test measures just one latent trait, all association between items can be explained by that trait; in other words, when the trait is partialed out, all items are uncorrelated (local stochastic independence). When more than one trait underlies a set of items, partialing out one of the traits or partialing out a composite of these traits will not result in vanishing associations. In other words, by inspecting the second-order frequencies, information is obtained with respect to violation of the axioms of local independence and unidimensionality.

Some authors (Hambleton & Swaminathan, 1985; Lord & Novick, 1968) do not recognize these axioms as separate ones. However, when a Rasch model with a learning parameter holds, instead of the simple Rasch model, we would argue that this is a unidimensional model, because the learning parameter is measured on the same latent continuum. When the simple Rasch model is applied to such a structure, local stochastic independence will prove to be violated, despite the fact that the data structure is unidimensional. So we would prefer to treat these assumptions as separate ones, though closely related.

The Q_2 statistic amounts to a comparison of observed and expected two-by-two contingency tables of item pairs. For items i and j a statistic can be obtained for each level group r:

$$q_{rij} = \frac{d^2}{E(n_{rij})} + \frac{d^2}{E(n_{r\bar{i}j})} + \frac{d^2}{E(n_{ri\bar{j}})} + \frac{d^2}{E(n_{r\bar{i}\bar{j}})} \tag{20}$$

where n_{rij} stands for the number of subjects having passed i and failed j, and so on for each of the four cells of the two-by-two table. The difference

$$d^2 = [n_{rij} - E(n_{rij})]^2 \tag{21}$$

is the same for each cell of the two-times-two table. The expectation of

n_{rij} is obtained by means of the conditional probability:

$$E(n_{rij}) = n_r \times \hat{\pi}_{rij} = n_r \times \frac{\hat{\varepsilon}_i \hat{\varepsilon}_j \times \hat{\gamma}_{r-2}^{(i,j)}}{\hat{\gamma}_r} \tag{22}$$

For each level group the item pairs can be totaled to obtain an overall statistic:

$$Q_{2r} = \frac{(k-3)}{(k-1)} \sum_{i=1}^{k-1} \sum_{j=i+1}^{k} q_{rij} \tag{23}$$

$(k - 3)/(k - 1)$ is a factor correcting for the interdependencies of the parameter estimates.

van den Wollenberg (1982c) proposed that there are theoretical arguments against the addition of Q_{2r} statistics to obtain an overall Q_2 statistic, but for all practical pruposes these objections seem rather trivial in their effect. It is even possible to obtain the Q_2 statistic for the unpartitioned sample, thus gaining in computing time and preventing the occurrence of extremely small expected frequencies, which can make the statistic unstable.

In numerous simulations, the Q_2 statistic proved to be a powerful tool in detecting multidimensionality where other statistics failed. In many applications a situation was encountered in which statistics of the Q_1 type were insignificant, whereas the Q_2 statistic was very significant indeed; the model violations could be (post hoc) very clearly interpreted in terms of more latent traits. The insensitivity of Q_1-type model tests for violation of dimensionality is thus not just a theoretical problem, but is frequently encountered in applications with real data.

4.3. The Splitter Item Technique

When two or more dimensions underly an item set, the items will be differentially interrelated: items that share the same dimension will be more strongly related than items measuring different dimensions. When an item is isolated from the test, its association with items of the same dimensional makeup will be stronger than with items that are different in this respect.

van den Wollenberg (1979, 1982a) made use of this fact in order to construct a procedure based on a Q_1-type statistic, which is sensitive for violation of unidimensionality.

Suppose, without loss of generality, that items of a test measure either trait A or trait B. Now an item of trait A is taken from the test and used as a splitting criterion: subjects having failed item A_i form one subsample, A^-; subjects having passed item A_i form the subsample A^+. Because of the differential relation of item A_i with A and B items, respectively, the two subsamples will show different patterns. Subsample A^+ consists on the average of subjects that are more able on A than the subjects of subsample A^-. So A items are relatively easy for the A^+ subsample and, therefore, relatively difficult for the A^- subsample. It is the relative difficulty of items over subsamples that is tested in Q_1-type model tests. This method was also reported to be successful by Molenaar (1983) and Formann (1981).

5. DESCRIPTIVE GOODNESS OF FIT

Using statistical model tests implies that one is investigating whether the model gives a (stochastically) perfect description of an empirical domain. However, in the social sciences one would be perfectly happy if a model gave a good description or approximation of the empirical domain, implying that one is ready to settle for a less than perfect model fit. This seems a realistic point of view, because the *a priori* probability of the model being true is zero.

A further point in this context is the fact that for application of latent trait models fair $(N = 400)$ to large samples are needed, which means that model tests have a higher power.

Because of the above points, it is our opinion that the evaluation of model fit should not only be based on statistical model tests. Descriptive goodness-of-fit statistics can make a contribution in addition to (but never instead of) model tests. It is our impression that descriptive goodness-of-fit statistics and procedures are used less than would be profitable.

5.1. THE REDUNDANCY INDEX

Martin-Lof (1974) introduced the redundancy index in order to have a goodness-of-fit measure confined into the range $(0, 1)$. The redundancy index is defined as the ratio of the logarithm of the likelihood ratio, and the logarithm of the total sample likelihood:

$$R = \frac{\ln \lambda}{L_t} \tag{24}$$

When the likelihood ratio approaches 1, the fit is good and the redundancy index approaches 0. The maximum of the log-likelihood ratio

TABLE 3
The Interpretation of the
Redundancy Index

Redundancy	Fit evaluation
1	Worst possible
0.1	Very bad
0.01	Bad
0.001	Good
0.0001	Very good

is equal to L_t and in that case redundancy is unity, representing the worst possible fit. Martin-Lof (1974) presents the calibration of the redundancy index shown in Table 3.

The redundancy index is only applicable to very large samples. Besides, it does not give information as to where the violations of the model are concentrated. Furthermore, the calibration of the measure is rather rude (good model fit is concentrated in the very small range (.001, 0.)). Finally, it only represents model fit in the sense of the Q_1-type statistics; violations of dimensionality are in principle not reflected in R.

All these points result in the measure not being used very frequently, and we would not advise otherwise.

5.2. GRAPHIC REPRESENTATION

When a sample is split up in two subsamples, and the item parameters are obtained in each subsample, a plot of the type shown in Figure 1 can be set up (e.g., Fischer, 1974). On the x axis the item parameter estimates of the first subsample are represented, whereas on the y axis the same is done for the second subsample. As the item parameter estimates should be equal in the subsamples the points representing the items should lie on the straight line at an angle of 45° through the origin. By means of this plot a global insight is gained regarding model fit, but information on the individual item level is also obtained. In the present representation, Item 6 can be identified as a possible violating item.

In the author's experience, this way of describing fit can be very helpful indeed, although it must be borne in mind that individual item points in the plot can have a different variability due to differences in statistical information. Again it should be stressed that only violation of sample independence is reflected in this graphic representation.

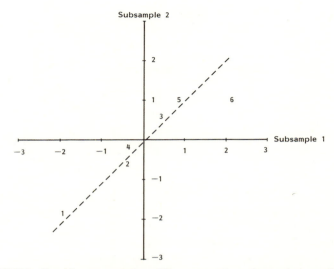

FIGURE 1. Graphic representation of item parameter estimates in two subsamples.

5.3. The Correlation between Parameters

As a complementary measure to the graphic representation, the correlation is often used. Because the item points in the plot should form an approximate straight line, the correlation should approach unity.

van den Wollenberg (1979) pointed out that a high correlation is a necessary condition for good descriptive model fit, but certainly not a sufficient condition. Correlation presupposes only a linear relation, whereas the Rasch model assumes an identity relation, which is more strict: the regression coefficients should both equal unity in addition to a perfect correlation in order to have a perfect model fit. The following auxiliary measure for use in connection with the correlation was suggested by van den Wollenberg (1979):

$$V_r = \frac{2(1 - r) \times s_{\hat{\sigma}} \times s_{\hat{\tau}}}{s_{(\hat{\sigma}-\hat{\tau})}^2} \tag{25}$$

where σ is used for the parameters of the first subsample and τ for the item parameters of the second subsample. The above measure can be interpreted as the proportion "badness-of-fit" variance that is reflected in imperfect correlation; its supplement to 1 can be interpreted as the proportion "badness-of-fit" variance that is due to unequal regression slopes.

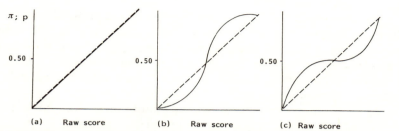

FIGURE 2. Probability plots for (a) an item that conforms to the model, (b) an item with too high discrimination power, and (c) an item with too low discrimination power.

5.4. PROBABILITY PLOTS AND PARAMETER PLOTS

Gustafsson (1979) uses probability plots in his PML program. When the sample is split up into several subsamples, according to raw score, a theoretical and an empirical probability for an item can be obtained in each of the subsamples, which together can be represented in the plots shown in Figure 2. The first plot represents the perfect situation. Probabilities based on the model (π) and observed relative frequencies (p) are in accordance. Over level groups (the x axis) the p values are equal to the model-based values and a straight line at an angle of 45° represents this perfect situation. The second plot shows an item with too high discrimination power; the last plot represents an item with too low discrimination.

Zwinderman and van den Wollenberg (in preparation) argue that these plots have the disadvantage that however large the deviations in the extreme parts of the continuum, the plot always shows a tendency toward the 45° degree line. They propose using the deviation of the subsample parameter estimate from the total group estimate, leading to plots of the type shown in Figure 3. The same situations as in Figure 2 are depicted

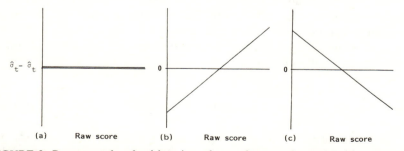

FIGURE 3. Parameter plots for (a) an item that conforms to the model, (b) an item with too high discrimination power, and (c) an item with too low discrimination power.

here. It can be observed that the disadvantage of the probability plots has vanished. In addition the deviations from the x axis can be simply interpreted as deviations of the subsample parameter estimate from the total sample estimate; in case of the p plots these deviations cannot even be compared on the same scale. All in all, parameter plots should be preferred to probability plots.

For all descriptive measures discussed up to now it can be said that they reflect violations of sample independence and are in principle insensitive to violation of the unidimensionality assumption.

5.5. Component Analysis of Standardized Q_2 Residuals

In the course of obtaining Q_2, residuals $n_{ij} - E(n_{ij})$ are evaluated. When the items i and j have more in common than the trait as determined in the analysis, there will be a positive residual association between i and j. In fact all items having the same residual trait in common will show positive residual associations. Zwinderman and van den Wollenberg (in preparation) used this notion in a heuristic method designed to detect Rasch homogeneous subsets of items. In this procedure the residuals are standardized to make association independent of p value; as a second step the principal components of the residual standardized association matrix are obtained. For example:

Of a set of ten items, five measure one trait, whereas the other items measure a second trait. The latent trait as determined in the analysis will typically be a weighted sum of the two input traits. Now the residuals of the first five items will be positively associated and the same will hold for the second set of five items. By a compensatory mechanism the residual associations between the two sets of items will be negative. In the principal component analysis two components will result, corresponding (possibly after rotation) with the two subsets of items. After identification of the subsets of items, a renewed Rasch analysis has to be performed to ascertain whether the subsets are Rasch homogeneous. In this sense component analysis of standardized residuals is only a heuristic method. The method has not yet been investigated thoroughly, but application on simulated data shows promise.

6. CONCLUSION

For evaluating fit to the Rasch model a variety of methods and procedures is available. In our opinion a thorough evaluation of model fit should contain the following elements:

- A test of parallelism of item characteristic curves (sample independence of raw score partitioning), by which the assumptions of monotonicity and sufficiency are investigated
- A test of unidimensionality, by which unidimensionality and local independence are investigated
- If feasible, a test on sample independence of partitions other than those based on raw score partitions, such as split-ups according to sex, age, and the like

These tests can be supplemented by descriptive and heuristic methods such as those described earlier.

Eventually evaluation of model fit and test design should always be a process of interaction between analytical results and substantial considerations.

7. REFERENCES

Allerup, P., & Sorber, G. (1977). *The Rasch model for questionnaires.* Copenhagen: Dansk Paedagogiske Institut.

Andersen, E. B. (1973a). *Conditional inference and models for measuring.* Unpublished doctoral dissertation, Mentalhygiejnisk Forsknings Institut, Copenhagen.

Andersen, E. B. (1973b). A goodness of fit test for the Rasch model. *Psychometrika, 38,* 123–140.

Fischer, G. H. (1974). *Einführung in die Theorie psychologischer Tests.* Bern: Huber.

Fischer, G. H., & Scheiblechner, H. H. (1970). Algorithmen und Programmen für das probabilistische Testmodell von Rasch. *Psychologische Beiträge, 12,* 23–51.

Formann, A. K. (1981). Über die Verwendung von Items als Teilungskriterium für Modellkontrollen im Modell von Rasch. *Zeitschrift für Experimentelle und Angewandte Psychologie, 28,* 541–560.

Glas, C. A. W. (1981). *Het Rasch model bij data in een onvolledig design.* Unpublished master's thesis, Utrecht.

Gustafsson, J. E. (1979). *PML: A computer program for conditional estimation and testing in the Rasch model for dichotomous items* (Report No. 85). Gothenburg: Institute of Education, University of Gothenburg.

Gustafsson, J. E. (1980). Testing and obtaining fit of data to the Rasch model. *British Journal of Mathematical and Statistical Psychology, 33,* 205–233.

Hambleton, R. K., & Swaminathan H. (1985). A look at psychometrics in The Netherlands. *Nederlands Tijdschrift voor de Psychologie, 40,* 446–451.

Kelderman, H. (1984). Loglinear Rasch model tests. *Psychometrika, 49,* 223–245.

Lord, F. M., & Novick, M. R. (1968). *Statistical theories of mental test scores.* Reading, MA: Addison-Wesley.

Martin-Lof, P. (1973). *Statistika Modeller, Anteckningar fran seminarier Lasaret 1969–1970 utarbetade av Rolf Sunberg obetydligt andrat nytryk, oktober 1973.* Stockholm: Institutet for Forsakringsmatematik och Matematisk vid Stockholms Universitet.

Martin-Lof, P. (1974). The notion of redundancy and its use as a quantitative measure of the discripancy between a statistical hypothesis and a set of observational data. *Scandinavian Journal of Statistics, 1,* 3–18.

Molenaar, I. W. (1983). Some improved diagnostics for failure of the Rasch model, *Psychometrika, 48,* 49–72.

Rasch, G. (1960). *Probabilistic models for some intelligence and attainment tests,* Kopenhagen: Nielsen & Lydiche.

Stelzl, I. (1979). Ist der Modelltest des Rasch-Modells geeignet, Homogenitatshypothesen zu prufen? Ein Bericht über Simulationsstudien mit inhomogenen Daten. *Zeitschrift für experimentelle und angewandte Psychologie, 26(4),* 652–672.

Wollenberg, A. L. van den (1979). *The Rasch model and time limit tests.* Unpublished doctoral dissertation, Studentenpers, Nijmegen.

Wollenberg, A. L. van den. (1980). On the Wright–Panchapakesan goodness of fit test for the Rasch model (Report 80MAOZ). Nijmegen: Katholieke Universiteit.

Wollenberg, A. L. van den. (1982a). A simple and effective method to test the dimensionality axiom of the Rasch model. *Applied Psychological Measurement, 6,* 83–91.

Wollenberg, A. L. van den. (1982b). Two new test statistics for the Rasch model. *Psychometrika, 47,* 123–139.

Wollenberg, A. L. van den. (1982c). On the applicability of the Q_2 test for the Rasch model. *Kwantitatieve Methoden, 5,* 30–55.

Wright, B. D., & Panchapakesan, N. A. (1969). A procedure for sample-free item analysis. *Educational and Psychological Measurement, 29,* 23–37.

Zwinderman, A. H., & Wollenberg, A. L. van den (in preparation). Heuristic procedures and descriptive measures for inspecting the fit of the Rasch model.

CHAPTER THREE

Latent Trait Models with Indicators of Mixed Measurement Level

GERHARD ARMINGER AND ULRICH KÜSTERS

1. INTRODUCTION

In this chapter we deal with the formulation and estimation of simultaneous equation models in metric latent endogenous variables that are connected to observed variables of any measurement level. The literature on this topic has focused on simultaneous equation models with metric (cf. Jöreskog & Sörbom, 1984) and ordinal indicators (Muthén, 1984). The use of ordinal indicators is based on an normal theory threshold concept implying an ordinal probit model. Concepts based on normal distribution theory are given up when qualitative variables are used as indicators for a latent metric variable (cf. the multinomial logit latent trait model of Bock, 1972). The proposed model structure consists of the following parts:

First, causal relations in metric latent variables are modeled as an ordinary simultaneous equation system including dependence on exogenous variables. Second, each observed variable is connected to a possibly vector-valued underlying variable—from now on denoted as the latent indicator—via threshold or random utility maximization submodels. Third, in the single indicator case the latent indicator is equal to an endogenous variable in the structural equation system. This may happen only if the latent indicator is not vector valued. Hence,

GERHARD ARMINGER AND ULRICH KÜSTERS • Department of Economics, University of Wuppertal, D-5600 Wuppertal 1, Federal Republic of Germany.

categorical indicators are excluded. Single indicators are limited to metric, censored metric, and ordinal variables (MCO variables). Fourth, in the multiple indicator case, several latent indicators of arbitrary measurement level are connected to one and only one latent endogenous variable by a factor analytic structure. Although it is in principle possible to consider models with two or more factors, only one-factor models are presented here. However, there may be more than one latent variable in the structural equation model for which multiple indicator sets exist.

Because the log-likelihood function of such a model involves multiple integrals, full information ML estimation is unfeasible. Similarly, because nonnormal distributions are used to formulate the model, estimation cannot be based on univariate and bivariate marginal likelihoods of the observed variables as utilized by Muthén (1984), although this estimation procedure may be employed for multiple MCO indicators. The estimation procedure proposed in this paper is based on the marginal likelihood corresponding to one endogenous variable of the simultaneous equation system.

2. MODEL FORMULATION

2.1. THE STRUCTURAL EQUATION SYSTEM

Let $\boldsymbol{\eta} \sim n \times 1$ be a vector of endogenous variables and $\mathbf{x} \sim p \times 1$ be a vector of exogenous variables related to each other in the structural equation model

$$\mathbf{B}\boldsymbol{\eta} = \boldsymbol{\Gamma}\mathbf{x} + \tilde{\boldsymbol{\varepsilon}} \tag{1}$$

with $\tilde{\boldsymbol{\varepsilon}} \sim N_n(\mathbf{O}, \tilde{\boldsymbol{\Omega}})$, $\mathbf{B}$ regular, and $\tilde{\boldsymbol{\Omega}}$ positive definite (cf. Schmidt, 1976). The reduced form of the system is

$$\boldsymbol{\eta} = \boldsymbol{\Pi}\mathbf{x} + \boldsymbol{\varepsilon} \tag{2}$$

with $\boldsymbol{\Pi} = \mathbf{B}^{-1}\boldsymbol{\Gamma}$ and $\boldsymbol{\varepsilon} = \mathbf{B}^{-1}\tilde{\boldsymbol{\varepsilon}} \sim N_n(\mathbf{O}, \boldsymbol{\Omega})$ with $\boldsymbol{\Omega} = \mathbf{B}^{-1}\tilde{\boldsymbol{\Omega}}\mathbf{B}'^{-1}$. The vector $\boldsymbol{\eta}$ is partitioned into $\boldsymbol{\eta}_1 \sim n_1 \times 1$ with single indicators of the MCO type and $\boldsymbol{\eta}_2 \sim n_2 \times 1$ with multiple indicators of any measurement level. For identification, the diagonal elements of $\boldsymbol{\Omega}$ corresponding to $\boldsymbol{\eta}_2$, that is $\omega_{n_1+1}^2, \ldots, \omega_n^2$, are set to 1.

2.2. MEASUREMENT RELATIONS FOR EQUATIONS WITH SINGLE INDICATORS

Let η_j be an element of $\boldsymbol{\eta}_1$. The corresponding single observed indicator Y_j is connected to η_j by classical threshold models (cf.

Amemiya, 1984; Bock, 1975; Maddala, 1983). We consider the following relations:

Y_j is *metric*. The measurement relation is the identity mapping

$$Y_j = \eta_j \tag{3}$$

Y_j is *censored metric,* that is, the values of the variable Y_j are observed only in a subset of the whole sample whenever the variable η_j is greater than a known threshold value τ_j. In the rest of the sample Y_j is set equal to τ_j, although η_j may vary from $-\infty$ to τ_j. (For a detailed discussion of the structure and of applications of the censored metric variable model, see Amemiya, 1984.) This measurement relation is called the tobit relation:

$$Y_j = \begin{cases} \tau_j & \text{if } \eta_j \le \tau_j \\ \eta_j & \text{if } \eta_j > \tau_j \end{cases} \tag{4}$$

Y_j is *ordinal* with c_j categories:

$$Y_j = k \qquad \text{iff } \tau_{j,k-1} < \eta_j \le \tau_{j,k}, \qquad k = 1, \ldots, c_j \tag{5}$$

with the restrictions

$$-\infty = \tau_{j,0} < \tau_{j,1} < \cdots < \tau_{j,c_j-1} < \tau_{j,c_j} = +\infty$$

The unknown thresholds are collected in the vector

$$\boldsymbol{\tau}_j \sim (c_j - 1) \times 1$$

The measurement relations are denoted by $Y_j = r_j(\eta_j)$. The inverse relations are denoted by $r_j^-(Y_j) = \{\eta_j : r_j(\eta_j) = Y_j\}$. Probability densities of Y_j are generated by the error term $\boldsymbol{\varepsilon}$ of the simultaneous equation system. The variance of η_j is ω_j^2 which is unrestricted in the metric and censored case and is set equal to 1 in the ordinal case for scale identification (Maddala & Lee, 1976; Nelson, 1976).

2.3. Factor Analytic Structure and Measurement Relations for Equations with Multiple Indicators

For each endogenous variable η_j that is an element of $\boldsymbol{\eta}_2$ there exists a set of multiple indicators $\{Y_{j,1}, \ldots, Y_{j,I_j}\}$ collected in the vector $\mathbf{Y}_j \sim I_j \times 1$. The sets of multiple indicators are disjoint. In this section we formulate models for the conditional density of $\mathbf{Y}_j$ given η_j. For the common conditional density $P(\mathbf{Y}_j \mid \eta_j)$ of a set, conditional (local) independence (Bartholomew, 1984) is assumed:

$$P(\mathbf{Y}_j \mid \eta_j) = \prod_{i=1}^{I_j} P(Y_{j,i} \mid \eta_j) \tag{6}$$

The conditional density $P(Y_{j,i} \mid \eta_j)$ is specified by the following relations:

First, to each $Y_{j,i}$ corresponds either a scalar latent indicator $Y_{j,i}^*$ if $Y_{j,i}$ is metric, censored metric, or ordinal, or a vector valued latent indicator $\mathbf{Y}_{j,i}^* \sim c_{j,i} \times 1$, if $Y_{j,i}$ is unordered categorical with $c_{j,i}$ categories. $Y_{j,i}$ and $\mathbf{Y}_{j,i}^*$ are connected by measurement relations described below.

Second, $\mathbf{Y}_{j,i}^*$ is connected with the corresponding latent endogenous variable η_j via a factor analytic model:

$$\mathbf{Y}_{j,i}^* = \boldsymbol{\alpha}_{j,i} + \boldsymbol{\lambda}_{j,i}\eta_j + \boldsymbol{\zeta}_{j,i} \tag{7}$$

$\boldsymbol{\alpha}_{j,i}$ is the regression constant in the regression of $\mathbf{Y}_{j,i}^*$ on η_j. $\boldsymbol{\lambda}_{j,i}$ is the vector of regression coefficients equivalent to factor loading in ordinary factor analysis. The error term $\boldsymbol{\zeta}_{j,i}$ is assumed to be independent of $\boldsymbol{\eta}$ and generates the stochastic variation of $Y_{j,i}$ given η_j.

Specification of the measurement relation between $Y_{j,i}$ and $\mathbf{Y}_{j,i}^*$ and of the term $\boldsymbol{\zeta}_{j,i}$ yields the conditional density $P(Y_{j,i} \mid \eta_j)$. The following combinations are considered:

$Y_{j,i}$ *metric:*

Measurement relation:

$$Y_{j,i} = Y_{j,i}^* \tag{8}$$

Error specification:

$$\zeta_{j,i} \sim N(0, \sigma_{j,i}^2) \tag{9}$$

Conditional density:

$$P(Y_{j,i} \mid \eta_j) = \phi(Y_{j,i} \mid \alpha_{j,i} + \lambda_{j,i}\eta_j, \sigma_{j,i}^2) \tag{10}$$

with

$$\phi(z \mid \mu, \sigma^2) = (2\pi\sigma^2)^{-1/2} \exp[-(1/2\sigma^2)(z - \mu)^2]$$

$Y_{j,i}$ *censored metric* with known threshold value $\tau_{j,i}$:

Measurement relation:

$$Y_{j,i} = \begin{cases} \tau_{j,i} & \text{if } Y_{j,i}^* \leq \tau_{j,i} \\ Y_{j,i}^* & \text{if } Y_{j,i}^* > \tau_{j,i} \end{cases} \tag{11}$$

Error specification:

$$\zeta_{j,i} \sim N(0, \sigma_{j,i}^2) \tag{12}$$

Conditional density (cf. Amemiya, 1984):

$$P(Y_{j,i} \mid \eta_j) = \begin{cases} \int\int_{-\infty}^{\tau_{j,i}} \phi(Y^* \mid \alpha_{j,i} + \lambda_{j,i}\eta_j, \sigma_{j,i}^2)\, dY^* & \text{if } Y_{j,i} = \tau_{j,i} \\ \phi(Y_{j,i} \mid \alpha_{j,i} + \lambda_{j,i}\eta_j, \sigma_{j,i}^2) & \text{if } Y_{j,i} > \tau_{j,i} \end{cases} \tag{13}$$

$Y_{j,i}$ *ordinal* with $c_{j,i}$ categories:

Measurement relation:

$$Y_{j,i} = k \qquad \text{iff } \tau_{j,i,k-1} < Y_{j,i}^* \leq \tau_{j,i,k}, \qquad k = 1, \ldots, c_{j,i} \tag{14}$$

with

$$-\infty = \tau_{j,i,0} < \tau_{j,i,1} \cdots < \tau_{j,i,c_{j,i}} = +\infty$$

Error specification:

$$\zeta_{j,i} \sim N(0, 1) \tag{15}$$

Conditional density (cf. McKelvey & Zavoina, 1975):

$$P(Y_{j,i} = k \mid \eta_j) = \int_{\tau_{j,i,k-1}}^{\tau_{j,i,k}} \phi(Y^* \mid \alpha_{j,i} + \lambda_{j,i}\eta_j, 1)\, dY^* \tag{16}$$

The variance in the error specification is set equal to 1 for the purpose of identification (Nelson, 1976). The normal density specification of the error term may be replaced by other densities. Examples are the logistic density yielding the ordinal logit model and the extreme value density yielding the complementary log(−log) ordinal model (McCullagh, 1980).

$Y_{j,i}$ *unordered categorical* with $c_{j,i}$ categories:

Measurement relation:

$$Y_{j,i} = k \qquad \text{iff } Y_{j,i,k}^* > Y_{j,i,l}^* \qquad \text{for } l = 1, \ldots, c_{j,i}, \qquad l \neq k \tag{17}$$

Error specification:

$$\zeta_{j,i,k}, \qquad k = 1, \ldots, c_{j,i}$$

are independently identically distributed with the extreme value distribution

$$F(\zeta_{j,i,k}) = \exp[-\exp(-\zeta_{j,i,k})] \tag{18}$$

yielding the multinomial logit (McFadden, 1974).
Conditional density:

$$P(Y_{j,i} = k \mid \eta_j) = \frac{\exp(\alpha_{j,i,k} + \lambda_{j,i,k}\eta_j)}{\sum_{l=1}^{c_{j,i}} \exp(\alpha_{j,i,l} + \lambda_{j,i,l}\eta_j)} \tag{19}$$

The first category serves as reference category. Hence, $\alpha_{j,i,1}$ and $\lambda_{j,i,1}$ are set to 0 for identification.

Again other multivariate error distributions for $\zeta_{j,i}$ may be employed. One example is the multivariate normal yielding conditional multinomial probits that are very cumbersome to compute (Daganzo, 1979). Another example is the generalized extreme value density yielding conditional tree extreme value models for preference trees (McFadden, 1978, 1981).

2.4. THE JOINT DENSITY OF EACH OBSERVATION

In addition to local independence of $(\mathbf{Y}_j \mid \eta_j)$ for each equation we assume local independence of $\mathbf{Y}_j$, $\mathbf{Y}_{j'}$, between equations. The single indicators and the sets of multiple indicators are collected in a column vector $\mathbf{Y} = (Y_1, \ldots, Y_{n_1}, \mathbf{Y}'_{n_1+1}, \ldots, \mathbf{Y}'_n)'$. The resulting joint density of $(\mathbf{Y} \mid \mathbf{x})$ for each observation is a mixture density with the multivariate normal as mixing density and the densities of Section 2.3 as conditional densities given $\boldsymbol{\eta}$:

$$P(\mathbf{Y} \mid \mathbf{x}) = \int_{r_1^-(Y_1)} \cdots \int_{r_{n_1}^-(Y_{n_1})} \underbrace{\int \cdots \int}_{\mathbb{R}^{n_2}} P(\mathbf{Y} \mid \boldsymbol{\eta})\phi(\boldsymbol{\eta} \mid \boldsymbol{\Pi}\mathbf{x}, \boldsymbol{\Omega}) \, d\boldsymbol{\eta} \tag{20}$$

where $\phi(\boldsymbol{\eta} \mid \boldsymbol{\Pi}\mathbf{x}, \boldsymbol{\Omega})$ is the multivariate normal density

$$\phi(\boldsymbol{\eta} \mid \boldsymbol{\Pi}\mathbf{x}, \boldsymbol{\Omega}) = (2\pi)^{-n/2} |\boldsymbol{\Omega}|^{-1/2} \exp[-\tfrac{1}{2}(\boldsymbol{\eta} - \boldsymbol{\Pi}\mathbf{x})'\boldsymbol{\Omega}^{-1}(\boldsymbol{\eta} - \boldsymbol{\Pi}\mathbf{x})] \tag{21}$$

with $|\Omega|$ as the determinant of Ω and where $P(\mathbf{Y} \mid \boldsymbol{\eta})$ is the conditional density of the indicators $\mathbf{Y}$ given the latent traits $\boldsymbol{\eta}$:

$$P(\mathbf{Y} \mid \boldsymbol{\eta}) = \prod_{j=n_1+1}^{n} \prod_{i=1}^{l_j} P(Y_{j,i} \mid \eta_j) \qquad (22)$$

If the inverse measurement relation $r_i^-(Y_i) = \{\eta_i\}$ is used, the corresponding integral is omitted.

A random sample of size T of independent drawings $\{\mathbf{Y}, \mathbf{x}\}_t$, $t = 1, \ldots, T$ is considered. The regressors $\mathbf{x}_t$ may either be fixed or stochastic depending on the nature of the process generating the data (experimental versus observational studies). However, the kernels of the log-likelihood function are equal (cf. Anderson & Philips, 1981) if the marginal density of $\mathbf{x}$ is not parametrized.

3. IDENTIFICATION PROBLEMS

3.1. FULL SYSTEM IDENTIFICATION

Criteria for the global identification in the sense of observationally equivalent structures with regard to all structural parameters

$\mathbf{B}, \boldsymbol{\Gamma}, \tilde{\boldsymbol{\Omega}}, \boldsymbol{\tau}_i, \quad i = 1, \ldots, n_1;$

$$((\alpha_{j,i}, \lambda_{j,i}, \tau_{j,i}, \Sigma_{j,i}), i = 1, \ldots, l_j), \qquad j = n_1 + 1, \ldots, n$$

are not known. The local identification status may in principle be checked by computing the rank of the information matrix based on the full likelihood [equation (20)] at the true parameter value (McDonald & Krane, 1977; Rothenberg, 1971). Because of the complexity of the likelihood function, even the check of the rank of the information matrix at the estimated parameter value is impossible in today's practice.

3.2. REDUCED FORM IDENTIFICATION FOR EACH EQUATION

Identification in the single indicator case is straightforward (Maddala & Lee, 1976). The only additional restriction required is that in ordinal probits $\tau_{i,1}$ is set to 0 if $\mathbf{x}$ contains a constant regressor.

Next we consider identification, if all of the multiple indicators are of the MCO type. In this case the marginal likelihood is given by

$$
P(\mathbf{Y}_j \mid \mathbf{x}) = \int_{\mathbb{R}} \int_{r_{j,1}(Y_{j,1})} \cdots \int_{r_{j,I_j}(Y_{j,I_j})} \tag{23}
$$
$$
\times \left\{ \prod_{i=1}^{I_j} \phi(Y_{j,i}^* \mid \alpha_{j,i} + \lambda_{j,i}\eta_j, \sigma_{j,i}^2)\phi(\eta_j \mid \mathbf{\Pi}_{j\cdot}\mathbf{x}, 1) \right\} d\mathbf{Y}_j^* \, d\eta_j
$$

which may be simplified by using conditional normal theory and integrating over $\mathbb{R}$ with respect to η_j. The following notation is employed:

$$
\boldsymbol{\alpha}_j = \begin{bmatrix} \alpha_{j,1} \\ \vdots \\ \alpha_{j,I_j} \end{bmatrix}, \qquad \boldsymbol{\lambda}_j = \begin{bmatrix} \lambda_{j,1} \\ \vdots \\ \lambda_{j,I_j} \end{bmatrix}, \qquad \boldsymbol{\Sigma}_j = \mathrm{diag}\{\sigma_{j,1}^2, \ldots, \sigma_{j,I_j}^2\} \tag{24}
$$

With this notation $P(\mathbf{Y}_j \mid \mathbf{x})$ may be written as

$$
P(\mathbf{Y}_j \mid \mathbf{x}) = \int_{r_{j,1}(Y_{j,1})} \cdots \int_{r_{j,I_j}(Y_{j,I_j})} \tag{25}
$$
$$
\times \phi(\mathbf{Y}_j^* \mid \boldsymbol{\alpha}_j + \boldsymbol{\lambda}_j \mathbf{\Pi}_{j\cdot}\mathbf{x}, \boldsymbol{\lambda}_j\boldsymbol{\lambda}_j' + \boldsymbol{\Sigma}_j) \, dY_j^*
$$

The covariance structure of the conditional density in the integral is identical with the covariance structure for a one factor model of classical factor analysis. Hence, only certain sign patterns in the correlation matrix are admissible under the model. Multiple indicators of the MCO type allow special estimation strategies that are more easily computed than the general estimation strategy described below. For this purpose, the univariate and bivariate marginal likelihoods for one or, respectively, two indicators can be used (see Muthén, 1984).

Returning to the identification problem, it follows that the underlying latent variables $Y_{j,i}^*$ are normally distributed with

$$
E(Y_{j,i}^* \mid \mathbf{x}) = \alpha_{j,i} + \lambda_{j,i}\mathbf{\Pi}_{j\cdot}\mathbf{x} \tag{26}
$$
$$
V(Y_{j,i}^* \mid \mathbf{x}) = \lambda_{j,i}^2 + \sigma_{j,i}^2 \tag{27}
$$

and

$$
\mathrm{cov}(Y_{j,i}^*, Y_{j,h}^* \mid \mathbf{x}) = \lambda_{j,i}\lambda_{j,h} \tag{28}
$$

In the case of metric or censored metric indicators, the variance

$V(Y^*_{j,i} \mid \mathbf{x})$ can be estimated from the data. If $Y_{j,i}$ is an ordinal indicator, the variance $V(Y^*_{j,i} \mid \mathbf{x})$ cannot be estimated from the data. Similarly to probit analysis and dichotomous factor analysis (Christoffersson, 1975; Muthén, 1978; Muthén & Christoffersson, 1981), a variance restriction must be imposed. Here $\sigma^2_{j,i}$ has been set to 1 [cf. equation (15)]. To be able to employ the usual probit convention of restricting the variance of the latent indicator to 1, we use the following one-to-one transformation of parameters, which will be estimated below:

$$\bar{\alpha}_{j,i} = \alpha_{j,i}/(\lambda^2_{j,i} + 1)^{1/2} \tag{29}$$

$$\bar{\lambda}_{j,i} = \lambda_{j,i}/(\lambda^2_{j,i} + 1)^{1/2} \tag{30}$$

The parameters $\mathbf{\Pi}_j$ are not changed by this transformation, as is seen from looking at the expectation and covariance structure of the corresponding transformed variable

$$\bar{Y}^*_{j,i} = Y^*_{j,i}/(\lambda^2_{j,i} + 1)^{1/2} \tag{31}$$

$$E(\bar{Y}^*_{j,i} \mid \mathbf{x}) = \bar{\alpha}_{j,i} + \bar{\lambda}_{j,i}\mathbf{\Pi}_j\mathbf{x} \tag{32}$$

$$V(\bar{Y}^*_{j,i} \mid \mathbf{x}) = 1$$

$$\mathrm{cov}(\bar{Y}^*_{j,i}, Y^*_{j,h}) = \bar{\lambda}_{j,i}\lambda_{j,h}, \qquad h \neq i \tag{33}$$

At least three indicators are necessary to identify $\lambda_{j,i}$ or, respectively, $\bar{\lambda}_{j,i}$ from the offdiagonals of the covariance matrix of the $Y^*_{j,i}$'s and the $\bar{Y}^*_{j,i}$'s. Given $\lambda_{j,i}$, $\sigma^2_{j,i}$ may be identified from equation (27) for metric and censored metric variables. Given, respectively, $\lambda_{j,i}$ or $\bar{\lambda}_{j,i}$, the parameters $\alpha_{j,i}$ or $\bar{\alpha}_{j,i}$ and $\mathbf{\Pi}_j$ may be identified from the expectation structure of equation (26) or (32). The first threshold $\tau_{j,i,1}$ is set to 0 for all ordinal indicators. If $\mathbf{x}$ does not contain a constant regressor, no further restrictions are necessary. If $\mathbf{x}$ contains a constant regressor, the additional restriction $\alpha_{j,1} = 0$ is used.

Now we consider the case of multiple indicators for one equation in general. Although we cannot directly transfer the analysis of the MCO case, we conjecture that the following conditions are sufficient:

- At least three indicators are employed.
- $\alpha_{j,1}$, if the first indicator is not categorical—or $\alpha_{j,1,2}$, if the first indicator is categorical—is set to 0.

Again, the local identification status can be checked by computing the rank of the information matrix of the reduced form for one equation. If a solution of the likelihood equation is obtained by using first derivatives of the marginal log-likelihood function, the information matrix may be approximated by its empirical counterpart [cf. equation (62)] on which the identification check may be performed.

3.3. IDENTIFICATION IN THE SIMULTANEOUS EQUATION SYSTEM

Identification conditions for the complete structural equation system are given by Schmidt (1976). However, we assume that each equation is identified by a normalization restriction ($B_{ii} = 1$) and at least $n - 1$ exclusion restrictions on the coefficients $\mathbf{B}_{i.}$ and $\mathbf{\Gamma}_{i.}$.

4. MARGINAL LIKELIHOOD ESTIMATION FOR EACH EQUATION

Throughout this section, the indicator set for one equation only is considered. Hence, the equation index j is omitted.

4.1. SINGLE INDICATOR ESTIMATION

Estimation of the single indicator case is based on maximizing the log-likelihood function corresponding to the latent endogenous variable η_j, $j = 1, \ldots, n_1$:

$$P(Y_j \mid \mathbf{x}) = \int_{r_j^-(Y_j)} \phi(\eta \mid \mathbf{\Pi}_{j.}\mathbf{x}, \omega_j^2) \, d\eta \qquad (34)$$

Computation is straightforward using ordinal probit (McKelvey & Zavoina, 1975), tobit (Amemiya, 1984), and ordinary metric regression (Schmidt, 1976) ML estimation.

4.2. MULTIPLE INDICATOR ESTIMATION

4.2.1. ML Estimation via a Solution of the Log-Likelihood Equation

The one-equation marginal density for a multiple indicator set is given by

$$P(\mathbf{Y} \mid \mathbf{x}) = \int_{\mathbb{R}} \prod_{i=1}^{I} P(Y_i \mid \eta, \tau_i, \alpha_i, \lambda_i, \mathbf{\Sigma}_i) \phi(\eta \mid \mathbf{\Pi}_{.}\mathbf{x}, 1) \, d\eta \qquad (35)$$

Before the log-likelihood function and its first derivatives are given, the following notation is introduced. Let δ_i be the indicator specific parameters depending on the measurement level of Y_i.

Y_i metric:

$$\delta_i = (\tilde{\delta}_i', \sigma_i^2)', \qquad \tilde{\delta}_i = (\alpha_i, \lambda_i)' \qquad (36)$$

Y_i metric censored:

$$\delta_i = (\tilde{\delta}_i', \sigma_i^2)', \qquad \tilde{\delta}_i = (\alpha_i, \lambda_i)' \qquad (37)$$

Y_i ordinal:

$$\delta_i = (\tilde{\delta}_i', \tau_{i,2}, \ldots, \tau_{i,c_i-1})', \qquad \tilde{\delta}_i = (\alpha_i, \lambda_i)' \qquad (38)$$

Y_i categorical:

$$\delta_i = (\tilde{\delta}_{i2}', \ldots, \tilde{\delta}_{ic_i}')', \qquad \tilde{\delta}_{ik} = (\alpha_{ik}, \lambda_{ik})' \qquad (39)$$

All parameters are collected in a vector

$$\vartheta = (\delta_1', \ldots, \delta_I', \Pi.)' = (\delta', \Pi.)'$$

Also a vector $\tilde{\eta} = (1, \eta)'$ is introduced.

The log-likelihood function for the whole sample is

$$l(\vartheta) = \sum_{t=1}^{T} \ln P(Y_t \mid x_t) \qquad (40)$$

The general structure of the first derivative of the log-likelihood function is

$$\frac{\partial}{\partial \vartheta} l(\vartheta) = \sum_{t=1}^{T} [P(Y_t \mid x_t)]^{-1} \frac{\partial}{\partial \vartheta} P(Y_t \mid x_t) \qquad (41)$$

with

$$\frac{\partial}{\partial \vartheta} P(Y_t \mid x_t) = \int_{\mathbb{R}} \frac{\partial}{\partial \vartheta} \left\{ \prod_{i=1}^{I} P(Y_{ti} \mid \eta, \delta_i) \phi(\eta \mid \Pi.x_t, 1) \right\} d\eta \qquad (42)$$

Abbreviating

$$\prod_{i=1}^{I} P(Y_{ti} \mid \eta, \delta_i) \phi(\eta \mid \Pi.x_t, 1)$$

by $P_t\phi_t$, we obtain

$$\frac{\partial}{\partial \boldsymbol{\vartheta}'} P_t\phi_t = \left\{ \left(\frac{\partial}{\partial \boldsymbol{\delta}_i'} P_t\phi_t\right)_{i=1,\ldots,I}, \frac{\partial}{\partial \boldsymbol{\Pi}} P_t\phi_t \right\} \tag{43}$$

The structure of the subvectors is given by

$$\frac{\partial}{\partial \boldsymbol{\delta}_i} P_t\phi_t = \left[\frac{\partial}{\partial \boldsymbol{\delta}_i} P(Y_{ti} \mid \eta, \boldsymbol{\delta}_i)\right]\phi(\eta \mid \boldsymbol{\Pi}\mathbf{x}_t, 1) \prod_{\substack{j=1 \\ j\neq i}}^{I} P(Y_{tj} \mid \eta, \boldsymbol{\delta}_j) \tag{44}$$

It should be noted that the derivatives with respect to those parameters in $\boldsymbol{\vartheta}$ that are fixed for identification purposes (cf. Section 3.2) are omitted. The individual components of $[\partial P(Y_{ti} \mid \eta, \boldsymbol{\delta}_i)/\partial\boldsymbol{\delta}_i]$ depend on the measurement level of Y_{ti}.

Y_{ti} metric:

$$P(Y_{ti} \mid \eta, \boldsymbol{\delta}_i) = \phi(Y_{ti} \mid \boldsymbol{\delta}_i'\tilde{\boldsymbol{\eta}}, \sigma_i^2) \equiv \phi_{ti} \tag{45}$$

$$\frac{\partial}{\partial\boldsymbol{\delta}_i} P(Y_{ti} \mid \eta, \boldsymbol{\delta}_i) = \frac{1}{\sigma_i^2}(Y_{ti} - \boldsymbol{\delta}_i'\tilde{\boldsymbol{\eta}})\tilde{\boldsymbol{\eta}}\phi_{ti} \tag{46}$$

$$\frac{\partial}{\partial\sigma_i^2} P(Y_{ti} \mid \eta, \boldsymbol{\delta}_i) = \left[\frac{1}{2\sigma_i^4}(Y_{ti} - \boldsymbol{\delta}_i'\tilde{\boldsymbol{\eta}})^2 - \frac{1}{2\sigma_i^2}\right]\phi_{ti} \tag{47}$$

Y_{ti} censored metric:

$$P(Y_{ti} \mid \eta, \boldsymbol{\delta}_i) = \begin{cases} \Phi\left(\dfrac{\tau_i - \boldsymbol{\delta}_i'\tilde{\boldsymbol{\eta}}}{\sigma_i}\right) & \text{for } Y_{ti} = \tau_i \\[2mm] \phi_{ti} & \text{for } Y_{ti} > \tau_i \end{cases} \tag{48}$$

$$\frac{\partial}{\partial\boldsymbol{\delta}_i} P(Y_{ti} \mid \eta, \boldsymbol{\delta}_i) = \begin{cases} \dfrac{\partial}{\partial\boldsymbol{\delta}_i} \Phi\left(\dfrac{\tau_i - \boldsymbol{\delta}_i'\tilde{\boldsymbol{\eta}}}{\sigma_i}\right) & \text{for } Y_{ti} = \tau_i \\[2mm] \dfrac{\partial}{\partial\boldsymbol{\delta}_i} \phi_{ti} & \text{for } Y_{ti} > \tau \end{cases} \tag{49}$$

where $\Phi(x)$ denotes the c.d.f. of the standard normal distribution. The derivatives $(\partial\phi_{ti}/\partial\boldsymbol{\delta}_i)$ are the same as in the metric case [cf. equations (46) and (47)]. In the censored case, we find

$$\frac{\partial}{\partial\boldsymbol{\delta}_i} \Phi\left(\frac{\tau_i - \boldsymbol{\delta}_i'\tilde{\boldsymbol{\eta}}}{\sigma_i}\right) = -\frac{1}{\sigma_i}\tilde{\boldsymbol{\eta}}\phi\left(\frac{\tau_i - \boldsymbol{\delta}_i'\tilde{\boldsymbol{\eta}}}{\sigma_i} \,\middle|\, 0, 1\right) \tag{50}$$

$$\frac{\partial}{\partial\sigma_i^2} \Phi\left(\frac{\tau_i - \boldsymbol{\delta}_i'\tilde{\boldsymbol{\eta}}}{\sigma_i}\right) = -\frac{1}{2(\sigma_i^2)^{3/2}}(\tau_i - \boldsymbol{\delta}_i'\tilde{\boldsymbol{\eta}})\phi\left(\frac{\tau_i - \boldsymbol{\delta}_i'\tilde{\boldsymbol{\eta}}}{\sigma_i} \,\middle|\, 0, 1\right) \tag{51}$$

Y_{ti} *ordinal:*

$$P(Y_{ti} \mid \eta, \boldsymbol{\delta}_i) = \Phi(\tau_{i,k} - \boldsymbol{\delta}_i'\bar{\boldsymbol{\eta}}) - \Phi(\tau_{i,k-1} - \boldsymbol{\delta}_i'\bar{\boldsymbol{\eta}}) \qquad \text{for } Y_{ti} = k \quad (52)$$

$$\frac{\partial}{\partial \boldsymbol{\delta}_i} P(Y_{ti} \mid \eta, \boldsymbol{\delta}_i) = -\{\phi(\tau_{i,k} - \boldsymbol{\delta}_i'\bar{\boldsymbol{\eta}} \mid 0, 1) - \phi(\tau_{i,k-1} - \boldsymbol{\delta}_i'\bar{\boldsymbol{\eta}} \mid 0, 1)\}\bar{\boldsymbol{\eta}} \tag{53}$$

$$\text{for } Y_{ti} = k$$

$$\frac{\partial}{\partial \tau_{i,k}} P(Y_{ti} \mid \eta, \boldsymbol{\delta}_i) = \begin{cases} \phi(\tau_{i,k} - \boldsymbol{\delta}_i'\bar{\boldsymbol{\eta}} \mid 0, 1) & \text{if } Y_{ti} = k; \quad k = 2, \ldots, c_i - 1 \\ -\phi(\tau_{i,k} - \boldsymbol{\delta}_i'\bar{\boldsymbol{\eta}} \mid 0, 1) & \text{if } Y_{ti} = k + 1 \\ 0 & \text{otherwise} \end{cases}$$
$$\tag{54}$$

In computing equations (52), (53), and (54), the restrictions $\tau_{i,0} = -\infty$, $\tau_{i,1} = 0$, $\tau_{i,c_i} = +\infty$ have to be taken into consideration.

Y_{ti} *categorical:*

$$P(Y_{ti} = k \mid \eta, \boldsymbol{\delta}_i) = \exp(\boldsymbol{\delta}_{ik}'\bar{\boldsymbol{\eta}}) \Big/ \left[1 + \sum_{l=2}^{c_i} \exp(\boldsymbol{\delta}_{il}'\bar{\boldsymbol{\eta}})\right] \tag{55}$$

$$\frac{\partial}{\partial \boldsymbol{\delta}_{ik}} P(Y_{ti} = k \mid \eta, \boldsymbol{\delta}_i) = [P(Y_{ti} = k \mid \eta, \boldsymbol{\delta}_i) - P(Y_{ti} = k \mid \eta, \boldsymbol{\delta}_i)^2]\bar{\boldsymbol{\eta}}, \tag{56}$$

$$k = 2, \ldots, c_i$$

$$\frac{\partial}{\partial \boldsymbol{\delta}_{ik}} P(Y_{ti} = l \mid \eta, \boldsymbol{\delta}_i) = -P(Y_{ti} = l \mid \eta, \boldsymbol{\delta}_i) P(Y_{ti} = k \mid \eta, \boldsymbol{\delta}_i)\bar{\boldsymbol{\eta}}, \tag{57}$$

$$l = 1, \ldots, c_i; \qquad k = 2, \ldots, c_i, \qquad k \neq l$$

The first derivatives of the log-likelihood with regard to $\boldsymbol{\Pi}'$ are

$$\frac{\partial}{\partial \boldsymbol{\Pi}'} P_t \phi_t = \left[\frac{\partial}{\partial \boldsymbol{\Pi}'} \phi(\eta \mid \boldsymbol{\Pi}.\mathbf{x}_t, 1)\right] \prod_{i=1}^{I} P(Y_{ti} \mid \eta, \boldsymbol{\delta}_i) \tag{58}$$

with

$$\frac{\partial}{\partial \boldsymbol{\Pi}'} \phi(\eta \mid \boldsymbol{\Pi}.\mathbf{x}_t, 1) = (\eta - \boldsymbol{\Pi}.\mathbf{x}_t)\mathbf{x}_t \phi(\eta \mid \boldsymbol{\Pi}.\mathbf{x}_t, 1) \tag{59}$$

Computing the first derivatives from equation (41) involves numerical integration of each term $P(\mathbf{Y}_t \mid \mathbf{x}_t)$ and $\partial P(\mathbf{Y}_t \mid \mathbf{x}_t)/\partial \vartheta$. Both terms involve $\phi(\eta \mid \boldsymbol{\Pi}.\mathbf{x}_t, 1)$ and are therefore of the form

$$\int_{-\infty}^{\infty} f(\eta) \exp[-\tfrac{1}{2}(\eta - \boldsymbol{\Pi}.\mathbf{x}_t)^2] \, d\eta = \int_{-\infty}^{\infty} f(2^{1/2}\xi + \boldsymbol{\Pi}.\mathbf{x}_t) \exp(-\xi^2) 2^{1/2} \, d\xi$$
$$\tag{60}$$

Using Gauss–Hermite integration with support points ξ_h and weights w_h, $h = 1, \ldots, H$ (Davis & Rabinowitz, 1984; Stroud & Secrest, 1966) yields approximately

$$\sum_{h=1}^{H} f(2^{1/2}\xi_h + \mathbf{\Pi}\mathbf{x}_t)2^{1/2}w_h \tag{61}$$

Estimates of the asymptotic covariance matrix $\mathbf{V}(\hat{\boldsymbol{\vartheta}})$ are usually computed using the inverse of the Fisher information matrix or the inverse of the observed information matrix. Since the second derivatives are cumbersome to compute, it is convenient to approximate the expected information matrix $\mathbf{I}(\boldsymbol{\vartheta})$ by its empirical counterpart, evaluated at the solution of the likelihood equations $\hat{\boldsymbol{\vartheta}}$:

$$\hat{\mathbf{I}}(\hat{\boldsymbol{\vartheta}}) \cong \frac{1}{T} \sum_{t=1}^{T} \frac{1}{P(\mathbf{Y}_t \mid \mathbf{x}_t)^2} \left[\frac{\partial P(\mathbf{Y}_t \mid \mathbf{x}_t)}{\partial \boldsymbol{\vartheta}} \right] \left[\frac{\partial P(\mathbf{Y}_t \mid \mathbf{x}_t)}{\partial \boldsymbol{\vartheta}'} \right] \Bigg|_{\vartheta = \hat{\vartheta}} \tag{62}$$

Given certain regularity conditions including global identifiability with respect to the $\boldsymbol{\vartheta}$ parametrization of the marginal density, the asymptotic distribution of $T^{1/2}(\hat{\boldsymbol{\vartheta}} - \boldsymbol{\vartheta})$ is $N(\mathbf{0}, \mathbf{I}(\boldsymbol{\vartheta})^{-1})$. Tests and confidence intervals for $\boldsymbol{\vartheta}$ may be based on this asymptotic result replacing $\mathbf{I}(\boldsymbol{\vartheta})$ by $\hat{\mathbf{I}}(\hat{\boldsymbol{\vartheta}})$. As computational procedures for solving the likelihood equation, algorithms employing only first derivatives such as the DFP, the BFGS algorithm (Luenberger, 1984), or a modified Fisher scoring method (Bock, 1972) may be used.

4.2.2. ML Estimation with the EM Algorithm

Basis for the application of the EM algorithm (Dempster, Laird, & Rubin, 1977) is the complete marginal density corresponding to the expression inside the integral of the incomplete marginal density (35)

$$P(\mathbf{Y}, \eta \mid \mathbf{x}, \boldsymbol{\vartheta}) = \prod_{i=1}^{I} P(Y_i \mid \eta, \boldsymbol{\delta}_i)\phi(\eta \mid \mathbf{\Pi}\mathbf{x}, 1) \tag{63}$$

In the EM algorithm the expected value of the logarithm of the complete data likelihood given the observed data $\{\mathbf{Y}, \mathbf{x}\}_t$, $t = 1, \ldots, T$ and the parameter estimates $\boldsymbol{\vartheta}^q$ from the previous qth iteration is maximized:

$$Q(\boldsymbol{\vartheta}^{q+1} \mid \boldsymbol{\vartheta}^q) = \sum_{t=1}^{T} E_\eta \{\ln P(\mathbf{Y}_t, \eta \mid \mathbf{x}_t, \boldsymbol{\vartheta}^{q+1}) \mid \mathbf{Y}_t, \boldsymbol{\vartheta}^q\} \tag{64}$$

The conditional density of $(\mathbf{Y}_t, \eta \mid \mathbf{x}_t)$ given $\mathbf{Y}_t$ and $\boldsymbol{\vartheta}^q$ is

$$g(\mathbf{Y}_t, \eta \mid \mathbf{x}_t, \mathbf{Y}_t, \boldsymbol{\vartheta}^q) = \frac{P(\mathbf{Y}_t \mid \eta, \boldsymbol{\delta}^q)\phi(\eta \mid \boldsymbol{\Pi}^q\mathbf{x}_t, 1)}{\int_{-\infty}^{\infty} P(\mathbf{Y}_t \mid \eta, \boldsymbol{\delta}^q)\phi(\eta \mid \boldsymbol{\Pi}^q\mathbf{x}_t, 1) \, d\eta} \tag{65}$$

Hence $Q(\boldsymbol{\vartheta}^{q+1} \mid \boldsymbol{\vartheta}^q)$ may be written as

$$\sum_{t=1}^{T} \int_{-\infty}^{\infty} \ln[P(\mathbf{Y}_t \mid \eta, \boldsymbol{\delta}^{q+1})\phi(\eta \mid \boldsymbol{\Pi}^{q+1}\mathbf{x}_t, 1)]g(\mathbf{Y}_t, \eta \mid \mathbf{x}_t, \mathbf{Y}_t, \boldsymbol{\vartheta}^q) \, d\eta \tag{66}$$

$$= \sum_{i=1}^{I} \sum_{t=1}^{T} \int_{-\infty}^{\infty} \ln[P(\mathbf{Y}_{ti} \mid \eta, \boldsymbol{\delta}_i^{q+1})]g(\mathbf{Y}_t, \eta \mid \mathbf{x}_t, \mathbf{Y}_t, \boldsymbol{\vartheta}^q) \, d\eta$$
$$+ \sum_{t=1}^{T} \int_{-\infty}^{\infty} \ln\phi(\eta \mid \boldsymbol{\Pi}^{q+1}\mathbf{x}_t, 1)g(\mathbf{Y}_t, \eta \mid \mathbf{x}_t, \mathbf{Y}_t, \boldsymbol{\vartheta}^q) \, d\eta \tag{67}$$

$$= \sum_{i=1}^{I} Q_i(\boldsymbol{\delta}_i^{q+1} \mid \boldsymbol{\vartheta}^q) + Q_0(\boldsymbol{\Pi}^{q+1} \mid \boldsymbol{\vartheta}^q) \tag{68}$$

As may be seen from the last line, we may maximize Q by optimizing $I + 1$ separate terms, each involving the posterior density $g(\mathbf{Y}_t, \eta \mid \mathbf{x}_t, \mathbf{Y}_t, \boldsymbol{\vartheta}^q)$. We must compute the denominator of this density—denoted by d_t—by numerical integration using equation (60). Again using numerical integration, each Q_i, $i = 1, \ldots, I$ may be written as

$$Q_i = \sum_{t=1}^{T} \frac{1}{d_t} \int_{-\infty}^{\infty} \ln[P(Y_{ti} \mid \eta, \boldsymbol{\delta}_i^{q+1})]P(\mathbf{Y}_t \mid \eta, \boldsymbol{\delta}^q)\phi(\eta \mid \boldsymbol{\Pi}\mathbf{x}_t, 1) \, d\eta \tag{69}$$

$$\cong \sum_{t=1}^{T} \frac{1}{d_t} \sum_{h=1}^{H} \ln[P(Y_{ti} \mid 2^{1/2}\xi_h + \boldsymbol{\Pi}^q\mathbf{x}_t, \boldsymbol{\delta}_i^{q+1})]$$
$$\times P(\mathbf{Y}_t \mid 2^{1/2}\xi_h + \boldsymbol{\Pi}^q\mathbf{x}_t, \boldsymbol{\delta}^q)w_h/\pi^{1/2} \tag{70}$$

The maximization of Q_i, $i = 1, \ldots, I$ with respect to $\boldsymbol{\delta}_i^{q+1}$ corresponds to a sequence of single optimization problems with $T \times H$ cases and known weights

$$v_{th} = [w_h/(d_t\pi^{1/2})]P(\mathbf{Y}_t \mid 2^{1/2}\xi_h + \boldsymbol{\Pi}^q\mathbf{x}_t, \boldsymbol{\delta}^q) \tag{71}$$

The single optimization problems are

• Ordinary regression
• Tobit regression (Amemiya, 1984)
• Ordinal probit regression (McKelvey & Zavoina, 1975)
• Multinomial logit regression (Bock, 1975)

with known weights. In each problem the regressor η is replaced by $2^{1/2}\xi_h + \mathbf{\Pi}^q_{\cdot}\mathbf{x}_t$.

Numerical integration is also applied to maximize Q_0:

$$Q_0 \cong \sum_{t=1}^{T} \sum_{h=1}^{H} \{\ln \phi(2^{1/2}\xi_h + \mathbf{\Pi}^q_{\cdot}\mathbf{x}_t \mid \mathbf{\Pi}^{q+1}_{\cdot}\mathbf{x}_t, 1)\}v_{th} \qquad (72)$$

In this case, η is the dependent variable and is accordingly replaced by $2^{1/2}\xi_h + \mathbf{\Pi}^q_{\cdot}\mathbf{x}_t$. Maximizing Q_0 is equivalent to a sequence of weighted regressions with $T \times H$ cases and known weights v_{th}.

A special case of the above procedure for dichotomous factor analysis has been developed by Bock and Aitkin (1981). A similar procedure for finite mixture models has been developed by Aitkin and Rubin (1985). An estimator of the observed information matrix (Efron & Hinkley, 1977) may in principle be obtained from the formulas of Louis (1982). However, this procedure is computationally cumbersome.

5. LATENT TRAIT SCORE ESTIMATION

The estimation of the score of the latent endogenous variable η_j for each case $t = 1, \ldots, T$ is based only on the indicators $\mathbf{Y}_j$. Hence, the equation index j is omitted throughout this section. It should be noted that the estimation of the latent trait score of the jth equation must—at least in theory—be based on all $\mathbf{Y}_j$ that correspond to latent traits influencing directly or indirectly the latent variable under consideration. For computational reasons we use the limited information of one equation only.

Under this restriction the η score of case t can be estimated in two ways. In either case we consider the posterior density $g(\mathbf{Y}_t, \eta \mid \mathbf{x}_t, \mathbf{Y}_t, \hat{\vartheta})$ of equation (65) evaluated at the final estimate $\hat{\vartheta}$. The first estimate is obtained as when computing factor scores, by considering η as a case specific parameter η_t. Maximization is straightforward but tedious using first derivatives and corresponding iterative optimization routines. A second estimate is based on the fact that η is a random variable (cf. Bartholomew, 1981) with density given in equation (65) conditional on $\mathbf{Y}_t, \mathbf{x}_t$ and the true parameter vector ϑ. If ϑ is replaced by $\hat{\vartheta}$, it is possible to estimate the latent score as the expected value of $(\eta \mid \mathbf{x}_t, \mathbf{Y}_t, \hat{\vartheta})$.

5.1. Score Estimation from Single Indicators

In the case of a single (metric, censored, or ordinal) indicator for the equation under consideration, the a posteriori expected value of η conditional on Y_t and $\mathbf{x}_t$ depends only on the measurement relation:

$$\hat{\eta}_t = E_\eta(\eta \mid \mathbf{x}_t, Y_t, \hat{\boldsymbol{\vartheta}}) = \int_{r^-(Y_t)} \eta \frac{\phi(\eta \mid \hat{\mathbf{\Pi}} \mathbf{x}_t, \hat{\omega}^2)}{\int_{r^-(Y_t)} \phi(\eta \mid \hat{\mathbf{\Pi}} \mathbf{x}_t, \hat{\omega}^2) \, d\eta} \, d\eta \qquad (73)$$

Evaluation of this integral yields expected values of the ordinary or doubly truncated normal (cf. Maddala, 1983). Because $\hat{\eta}_t$ is conditional on $\mathbf{x}_t$, the $\hat{\eta}_t$'s are not identically distributed.

To derive the variance of $\hat{\eta}_t$, we first compute the expected value of η_t—conditional on $\mathbf{x}_t$ only—by integrating out Y at the true parameter value $\boldsymbol{\vartheta}$:

$$E_y(E_\eta(\eta \mid Y, \mathbf{x}_t, \boldsymbol{\vartheta})) = \mathbf{\Pi} \mathbf{x}_t \qquad (74)$$

Hence, the variance is

$$V(\hat{\eta}_t) = V_y(E_\eta(\eta \mid Y, \mathbf{x}_t, \boldsymbol{\vartheta})) = E_y[E_\eta(\eta \mid Y, \mathbf{x}_t, \boldsymbol{\vartheta}) - \mathbf{\Pi} \mathbf{x}_t]^2$$
$$= \int_{D(Y)} [E_\eta(\eta \mid Y, \mathbf{x}_t, \boldsymbol{\vartheta}) - \mathbf{\Pi} \mathbf{x}_t]^2 P(Y \mid \mathbf{x}_t, \boldsymbol{\vartheta}) \, dY \qquad (75)$$

where $D(Y)$ is the domain of Y and the integral is replaced by a sum, if Y_i is discrete.

5.2. Score Estimation from Multiple Indicators

In the case of multiple indicators, the latent trait score is estimated by

$$\hat{\eta}_t = E_\eta(\eta \mid \mathbf{x}_t, \mathbf{Y}_t, \hat{\boldsymbol{\vartheta}}) = \int_{\mathbb{R}} \eta \, g(\mathbf{Y}_t, \eta \mid \mathbf{x}_t, \mathbf{Y}_t, \hat{\boldsymbol{\vartheta}}) \, d\eta \qquad (76)$$

Using numerical integration [cf. equation (70)] we approximate the integral by

$$\hat{\eta}_t \cong \sum_{h=1}^{H} (2^{1/2}\xi_h + \mathbf{\Pi} \mathbf{x}_t)\hat{v}_{th} \qquad (77)$$

where $\hat{v}_{th}$ is equal to v_{th} evaluated at $\hat{\boldsymbol{\vartheta}}$.

The variance is derived similarly to the single indicator case:

$$V(\hat{\eta}_t) = V_y(E_\eta(\eta \mid \mathbf{Y}, \mathbf{x}_t, \boldsymbol{\vartheta})) = E_y[E_\eta(\eta \mid \mathbf{Y}, \mathbf{x}_t, \boldsymbol{\vartheta}) - \boldsymbol{\Pi}.\mathbf{x}_t]^2$$

$$= \int_{D(Y_1)} \cdots \int_{D(Y_l)} [E_\eta(\eta \mid \mathbf{Y}, \mathbf{x}_t, \boldsymbol{\vartheta}) - \boldsymbol{\Pi}.\mathbf{x}_t]^2 P(\mathbf{Y} \mid \mathbf{x}_t) \, d\mathbf{Y} \tag{78}$$

Again $D(Y_i)$ is the domain of Y_i and an integral is replaced by a sum if Y_i is discrete.

The numerical computation of the variance involves integration and or summation over the whole domain $D(\mathbf{Y}) = \times_{i=1}^{I} D(Y_i)$ and is therefore hardly feasible for indicator sets of even modest size.

6. ESTIMATION OF STRUCTURAL PARAMETERS FROM THE REDUCED FORM PARAMETERS

In deriving estimates of structural parameters we follow Amemiya (1978b, 1979). For simplicity's sake we consider without loss of generality only the estimation of the structural parameters $\mathbf{B}_{1.}$ and $\boldsymbol{\Gamma}_{1.}$ of the first equation. According to the restrictions on $\mathbf{B}_{1.}$ and $\boldsymbol{\Gamma}_{1.}$ necessary for identification, the coefficients in $\mathbf{B}_{1.}$ and $\boldsymbol{\Gamma}_{1.}$ are arranged in the following pattern:

$$\mathbf{B}_{1.} = (1, \tilde{\mathbf{B}}_{1.}, \mathbf{0}) \sim 1 \times [1 + m_1 + (n - m_1 - 1)] \tag{79}$$

$$\boldsymbol{\Gamma}_{1.} = (\tilde{\boldsymbol{\Gamma}}_{1.}, \mathbf{0}) \sim 1 \times [p_1 + (p - p_1)] \tag{80}$$

The following partition of $\boldsymbol{\Pi}$ is of the same pattern as the inclusion and exclusion of variables in $\mathbf{B}_{1.}$:

$$\boldsymbol{\Pi} = \begin{cases} \boldsymbol{\Pi}_{1.} \sim 1 \times p \\ \tilde{\boldsymbol{\Pi}}^{(1)} \sim m_1 \times p \\ \tilde{\boldsymbol{\Pi}}^{(2)} \sim (n - m_1 - 1) \times p \end{cases} \tag{81}$$

Hence, the relation $\boldsymbol{\Gamma} = \mathbf{B}\boldsymbol{\Pi}$ may be written for the first equation as

$$(\tilde{\boldsymbol{\Gamma}}_{1.}, \mathbf{0}) = \boldsymbol{\Pi}_{1.} + \tilde{\mathbf{B}}_{1.}\tilde{\boldsymbol{\Pi}}^{(1)} \tag{82}$$

Replacing $\boldsymbol{\Pi}_1$ and $\tilde{\boldsymbol{\Pi}}^{(1)}$ by $\boldsymbol{\Pi}_{1.} + \hat{\boldsymbol{\Pi}}_{1.} - \hat{\boldsymbol{\Pi}}_{1.}$ and $\tilde{\boldsymbol{\Pi}}^{(1)} + \hat{\tilde{\boldsymbol{\Pi}}}^{(1)} - \hat{\tilde{\boldsymbol{\Pi}}}^{(1)}$ in the last equation yields

$$\hat{\boldsymbol{\Pi}}_{1.} = (\tilde{\boldsymbol{\Gamma}}_{1.}, \mathbf{0}) + \tilde{\mathbf{B}}_{1.}(-\hat{\tilde{\boldsymbol{\Pi}}}^{(1)}) + \boldsymbol{\delta} \tag{83}$$

with

$$\boldsymbol{\delta} = \hat{\boldsymbol{\Pi}}_{1.} - \boldsymbol{\Pi}_{1.} + \tilde{\mathbf{B}}_{1.}(\hat{\boldsymbol{\Pi}}^{(1)} - \tilde{\boldsymbol{\Pi}}^{(1)}) \tag{84}$$

This is equivalent to

$$\hat{\boldsymbol{\Pi}}_{1.} = \mathbf{bZ} + \boldsymbol{\delta}$$

with

$$\mathbf{b} = (\tilde{\mathbf{B}}_{1.}, \tilde{\boldsymbol{\Gamma}}_{1.}) \quad \text{and} \quad \mathbf{Z} = \begin{bmatrix} -\hat{\boldsymbol{\Pi}}^{(1)} \\ \hline \mathbf{I}_{p_1} \mid \mathbf{0} \end{bmatrix} \tag{85}$$

Straightforward least squares yields the estimator

$$\hat{\mathbf{b}} = \hat{\boldsymbol{\Pi}}_{1.} \mathbf{Z}'(\mathbf{ZZ}')^{-1} \tag{86}$$

Now we derive an estimate of the covariance matrix of $\hat{\mathbf{b}}$. For this purpose, we first approximate the ML estimate of the asymptotic joint covariance matrix of

$$\text{vec } \hat{\boldsymbol{\Pi}} = \begin{bmatrix} \hat{\boldsymbol{\Pi}}'_{1.} \\ \vdots \\ \hat{\boldsymbol{\Pi}}'_{n.} \end{bmatrix} \tag{87}$$

In a manner analogous to the derivation of the asymptotic distribution of the pseudo ML estimators for discrete choice models under stratified sampling for the endogenous variable (Manski & McFadden, 1981) it may be shown that $T^{1/2}(\text{vec } \hat{\boldsymbol{\Pi}} - \text{vec } \boldsymbol{\Pi}_0)$ is asymptotically normally distributed with expectation $\mathbf{0}$ and a covariance matrix with the structure $\mathbf{J}^{-1}\mathbf{A}\mathbf{J}^{-1\prime}$. (For a corresponding two-step estimator cf. Amemiya, 1978.) $\boldsymbol{\Pi}_0$ denotes the true parameter value of $\boldsymbol{\Pi}$. To avoid the rather complicated derivations of $\mathbf{J}$ and $\mathbf{A}$ ($\mathbf{J}_0$ and $\mathbf{A}_0$ denote these matrices at the true parameter value), only the approximation of the submatrices of $\mathbf{J}$ and $\mathbf{A}$ at the estimated parameter value for equation i and j needed for computation are given in the sequel. The submatrices of $\hat{\mathbf{J}}$ and $\hat{\mathbf{A}}$ corresponding to $\hat{\boldsymbol{\Pi}}_{i.}, \hat{\boldsymbol{\Pi}}_{j.}$ are denoted by $\mathbf{J}_{ij}$ and $\hat{\mathbf{A}}_{ij}$, $i, j = 1, \ldots, m$ and are computed by

$$\hat{\mathbf{A}}_{ij} = \frac{1}{T}\sum_{t=1}^{T} \frac{\partial \ln P(\mathbf{Y}_{ti} \mid \mathbf{x}_t)}{\partial \boldsymbol{\Pi}'_{i.}} \frac{\partial \ln P(\mathbf{Y}_{tj} \mid \mathbf{x}_t)}{\partial \boldsymbol{\Pi}_{j.}} \tag{88}$$

evaluated at the estimated parameter values $(\hat{a}_i, \hat{\lambda}_i, \hat{\tau}_i, \hat{\sigma}_i^2, \hat{\mathbf{\Pi}}_{i.}, \hat{\omega}_i)$ and

$$\hat{\mathbf{J}}_{ij} = \begin{cases} -\hat{\mathbf{A}}_{ii} & \text{for } i = j \\ \mathbf{0} & \text{for } i \neq j \end{cases} \tag{89}$$

This results in the estimates of the joint covariance matrix of $T^{1/2} (\text{vec } \hat{\mathbf{\Pi}} - \text{vec } \mathbf{\Pi}_0)$:

$$\hat{\mathbf{V}}_{ij} = (\hat{\mathbf{J}}^{-1} \hat{\mathbf{A}} \hat{\mathbf{J}}^{-1'})_{ij} = \begin{cases} \hat{\mathbf{A}}_{ii}^{-1} & \text{for } i = j \\ \hat{\mathbf{J}}_{ii}^{-1} \hat{\mathbf{A}}_{ij} \hat{\mathbf{J}}_{jj}^{-1} & \text{for } i \neq j \end{cases} \tag{90}$$

Hence vec $\hat{\mathbf{\Pi}}$ is asymptotically distributed as

$$\text{vec } \hat{\mathbf{\Pi}} \sim N(\text{vec } \mathbf{\Pi}_0, (1/T) \hat{\mathbf{V}}) \tag{91}$$

The asymptotic distribution of $T^{1/2}(\hat{\mathbf{b}} - \mathbf{b})$ may now be derived as follows. Replacing $\hat{\mathbf{\Pi}}_{1.}$ in equation (86) by $(\mathbf{bZ} + \boldsymbol{\delta})$ yields

$$T^{1/2}(\hat{\mathbf{b}} - \mathbf{b}) = T^{1/2} \boldsymbol{\delta} \mathbf{Z}'(\mathbf{ZZ}')^{-1} \tag{92}$$

Let $\hat{\mathbf{\Pi}}_a^{(1)}$ denote the first p_1 columns of $\tilde{\mathbf{\Pi}}^{(1)}$ and $\tilde{\mathbf{\Pi}}_b^{(1)}$ denote the rest of $\hat{\mathbf{\Pi}}^{(1)}$. From the consistency of $\hat{\mathbf{\Pi}}$ follows

$$\mathbf{C}_0 = p \lim \mathbf{Z}'(\mathbf{ZZ}')^{-1} = \left[-\tilde{\mathbf{\Pi}}^{(1)'} \;\middle|\; \frac{\mathbf{I}_{p_1}}{\mathbf{0}} \right] \left[\frac{\tilde{\mathbf{\Pi}}^{(1)} \tilde{\mathbf{\Pi}}^{(1)'} \;\middle|\; -\tilde{\mathbf{\Pi}}_a^{(1)}}{-\tilde{\mathbf{\Pi}}_a^{(1)'} \;\middle|\; \mathbf{I}_{p_1}} \right]^{-1}$$

With $B_{11} = 1$, the column vector $T^{1/2} \boldsymbol{\delta}'$ may be written as

$$T^{1/2} \boldsymbol{\delta}' = \left[\sum_{j=1}^{m_1+1} \mathbf{B}_{1j} T^{1/2} (\hat{\mathbf{\Pi}}_{j.} - \mathbf{\Pi}_{j.0}) \right]'$$

or

$$T^{1/2} \boldsymbol{\delta}' = \mathbf{D}_0 T^{1/2} \begin{bmatrix} \hat{\mathbf{\Pi}}_{1.}' - \mathbf{\Pi}_{1.}' \\ \vdots \\ \hat{\mathbf{\Pi}}_{m_1+1.}' - \mathbf{\Pi}_{m_1+1.}' \end{bmatrix}$$

with

$$\mathbf{D}_0 = (1, \tilde{\mathbf{B}}_{1.}) \otimes \mathbf{1}_p$$

Let $\mathbf{V}_0$ be the submatrix of $\mathbf{V}$ with the covariance matrices $\mathbf{V}_{ij}$ according to $\hat{\mathbf{\Pi}}_{i.}$, $\hat{\mathbf{\Pi}}_{j.}$, $i, j = 1, \ldots, m_1 + 1$. Then

$$T^{1/2}(\hat{\mathbf{b}} - \mathbf{b})' \overset{A}{\sim} N(\mathbf{0}, \mathbf{C}_0'\mathbf{D}_0\mathbf{V}_0\mathbf{D}_0'\mathbf{C}_0) \tag{93}$$

where $\overset{A}{\sim}$ denotes asymptotic equivalence.

When the asymptotic covariance matrix is estimated, the true parameters involved are replaced by their consistent estimates. An alternative derivation of $\text{Var}(\hat{\mathbf{b}})$ may be achieved by the application of the multivariate δ method (Serfling, 1980) again using the covariance matrix of $\hat{\mathbf{\Pi}}$.

The generalized least squares estimator of $\hat{\mathbf{b}}$ is obtained by employing the covariance matrix of $T^{1/2}\hat{\boldsymbol{\delta}}'$. This results in

$$\hat{\mathbf{b}}_{\text{GLS}} = \hat{\mathbf{\Pi}}_{1.}(\mathbf{D}_0\mathbf{V}_0\mathbf{D}_0')^{-1}\mathbf{Z}'[\mathbf{Z}(\mathbf{D}_0\mathbf{V}_0\mathbf{D}_0')^{-1}\mathbf{Z}']^{-1} \tag{94}$$

where the true parameters are replaced by their estimates. It should be noted that the matrix $\mathbf{D}_0$ involves the parameter $\tilde{\mathbf{B}}_{1.}$ Hence $\tilde{\mathbf{B}}_{1.}$ must be estimated by least squares using equation (86). The asymptotic distribution is given by

$$T^{1/2}(\hat{\mathbf{b}}_{\text{GLS}} - \mathbf{b})' \overset{A}{\sim} N(\mathbf{0}, [\mathbf{Z}(\mathbf{D}_0\mathbf{V}_0\mathbf{D}_0')^{-1}\mathbf{Z}']^{-1}) \tag{95}$$

7. EXTENSIONS

The marginal estimation strategy described above may be extended in a straightforward but tedious manner to estimate the correlation between two latent variables η_i and η_j using the marginal likelihood of the corresponding indicator sets. The case of two single indicators given exogenous variables yields a generalization of the polychoric (Olsson, Drasgow, & Dorans, 1982) correlation coefficient discussed in detail by Küsters (1987). The estimation strategy of the correlation between latent variables with multiple categorical indicators proposed by Andersen (1985) using partial marginal likelihood estimation may be extended to include indicators of mixed measurement level, simple indicator models, and exogeneous variables. Model formulation and details of the estimation procedure are given by Arminger and Küsters (1985). A crucial assumption of the estimation strategy discussed above is the conditional independence of indicators. This assumption is usually violated if panel data are analyzed where serial dependence between identical indicators measured at successive panel waves is supposed. A discussion of this issue can be found in Arminger and Küsters (1987).

8. REFERENCES

Aitkin, M., & Rubin, D. B. (1985). Estimation and hypothesis testing in finite mixture models. *Journal of the Royal Statistical Society, B47,* 67–75.

Amemiya, T. (1978a). On a two-step estimation of a multivariate logit model. *Journal of Econometrics, 8,* 13–21.

Amemiya, T. (1978b). The estimation of a simultaneous equation generalized probit model. *Econometrica, 46,* 1193–1205.

Amemiya, T. (1979). The estimation of a simultaneous equation tobit model. *International Economic Review, 20,* 169–181.

Amemiya, T. (1984). Tobit models: A survey. *Journal of Econometrics, 24,* 3–61.

Andersen, E. B. (1985). Estimating latent correlation between repeated testings. *Psychometrika, 50,* 3–16.

Anderson, J. A., & Philips, P. R. (1981). Regression, discrimination and measurement models of ordered categorical variables. *Applied Statistics, 30,* 22–31.

Arminger, G., & Küsters, U. (1985). *Latent trait and correlation models with indicators of mixed measurement level.* Preliminary manuscript prepared for the Thirteenth Symposium on Latent Trait and Latent Class Models in Educational Research, IPN Kiel.

Arminger, G., & Küsters, U. (1987). *New developments in latent variable measurement and structural equation models.* Unpublished manuscript, University of Wuppertal.

Bartholomew, D. J. (1981). Posterior analysis of the factor model. *British Journal of Mathematical and Statistical Psychology, 34,* 93–99.

Bartholomew, D. J. (1984). The foundations of factor analysis. *Biometrika, 71,* 221–232.

Bock, R. D. (1972). Estimating item parameters and latent ability when responses are scored in two or more nominal categories. *Psychometrika, 37,* 29–51.

Bock, R. D. (1975). *Multivariate statistical methods in behavioral research.* New York: McGraw-Hill.

Bock, R. D., & Aitkin, M. (1981). Marginal maximum likelihood estimation of item parameters: Application of an EM-algorithm. *Psychometrika, 40,* 443–459.

Christoffersson, A. (1975). Factor analysis of dichotomized variables. *Psychometrika, 40,* 5–32.

Daganzo, C. (1979). *Multinomial probit—The theory and its application to demand forecasting.* New York: Academic Press.

Davis, P. J., & Rabinovitz, P. (1984). *Methods of numerical integration.* Orlando: Academic Press.

Dempster, A. P., Laird, N. M., & Rubin, D. B. (1977). Maximum likelihood of incomplete data via the EM algorithm (with discussion). *Journal of the Royal Statistical Society, B39,* 1–38.

Efron, B., & Hinkley, D. V. (1978). Assessing the accuracy of the maximum likelihood estimator: Observed versus expected Fisher information. *Biometrika, 65,* 475–487.

Jöreskog, K. G., & Sörbom, D. (1984). LISREL VI—*Analysis of linear structural relationships by maximum likelihood, instrumental variables and least squares methods.* Mooresville, Indiana: Scientific Software.

Küsters, U. (1987). *Hierarchische Mittelwert- und Kovarianzstrukturmodelle mit nichtmetrischen endogenen Variablen.* Heidelberg: Physica-Verlag.

Louis, T. A. (1982). Finding the observed information matrix when using the EM-algorithm. *Journal of the Royal Statistical Society, B44*, 226–233.

Luenberger, D. G. (1984). *Linear and nonlinear programming.* Reading, Massachusetts: Addison-Wesley.

Maddala, G. S. (1983). *Limited-dependent and qualitative variables in econometrics.* Cambridge: Cambridge University Press.

Maddala, G. S., & Lee, L. F. (1976). Recursive models with qualitative endogenous variables. *Annals of Economic and Social Measurement, 5/4*, 525–545.

Manski, C. F., & McFadden, D. (1981). Alternative estimators and sample designs for discrete choice analysis. In C. F. Manski & D. McFadden (Eds.), *Structural analysis of discrete data with econometric applications* (pp. 2–50). Cambridge, Massachusetts: MIT Press.

McCullagh, P. (1980). Regression models for ordinal data (with discussion). *Journal of the Royal Statistical Society, B2*, 109–142.

McDonald, R. P., & Krane, W. R. (1977). A note on local identifiability and degrees of freedom in the asymptotic likelihood ratio test. *British Journal of Mathematical and Statistical Psychology, 30*, 198–203.

McFadden, D. (1974). Conditional logit analysis of qualitative choice behavior. In P. Zarembka (Ed.), *Frontiers in econometrics* (pp. 105–142). New York: Academic Press.

McFadden, D. (1978). Modeling the choice of residential location. In A. Karlgvist (Ed.), *Spatial interaction theory and residential location* (pp. 75–96). Amsterdam: North-Holland.

McFadden, D. (1981). Econometric models of probabilistic choice. In C. F. Manski & D. McFadden (Eds.), *Structural analysis of discrete data with econometric applications* (pp. 198–272). Cambridge, Massachusetts: MIT Press.

McKelvey, R. D., & Zavoina, W. (1975). A statistical model for the analysis of ordinal level dependent variables. *Journal of Mathematical Sociology, 4*, 103–120.

Muthén, B. (1978). Contributions to factor analysis of dichotomous variables. *Psychometrika, 43*, 551–560.

Muthén, B. (1984). A general structural equation model with dichotomous, ordered categorical, and continuous latent variable indicators. *Psychometrika, 49*, 115–132.

Muthén, B., & Christoffersson, A. (1981). Simultaneous factor analysis of dichotomous variables in several groups. *Psychometrika, 46*, 407–419.

Nelson, F. D. (1976). On a general computer algorithm for the analysis of models with limited dependent variables. *Annals of Economic and Social Measurement, 5*, 493–509.

Olsson, U., Drasgow, F., & Dorans, N. J. (1982). The polyserial correlation coefficient. *Psychometrika, 47*, 337–347.

Rothenberg, T. J. (1971). Identification in parametric models. *Econometrica, 39*, 577–591.

Schmidt, P. (1976). *Econometrics.* New York: Marcel Dekker.

Serfling, R. J. (1980). *Approximation theorems of mathematical statistics.* New York: Wiley.

Stroud, A., & Secrest, D. (1966). *Gaussian quadrature formulas.* Englewood Cliffs, New Jersey: Prentice Hall.

PART II

LATENT CLASS THEORY

CHAPTER FOUR

New Developments in Latent Class Theory

ROLF LANGEHEINE

1. INTRODUCTION

A feature common to most models considered in the first part of this book may be easily depicted by Figure 1. In latent trait models, the point of departure is a set of manifest categorical variables (indicators, items, say A, B, and C), which may be related to each other in some way (cf. the curved lines). The crucial assumption now is that these relationships are conceived as being due to some *continuous latent variable X*. That is, if the model holds, the relationships between the manifest variables will vanish and the structure will be depicted by the straight lines going from X to A, B, and C. Several people, however, have questioned whether this procedure is advisable in all instances. Latent trait models strive for a relatively sophisticated scaling property of the latent variable (most models aim at least at an interval scale) which often remains unused for subsequent interpretation of the data. In fact, we are often simply interested in certain groups or types of persons (see Rost, Chapter 7, this volume), that means that we need no more than a *categorical or nominal latent variable*. This is exactly what latent class models assume.

The basic conceptualization and assumptions of latent class models (LCM) or latent class analysis (LCA) may be outlined very briefly. A, B, and C are categorical manifest variables[1] and there is some relationship

---

[1] To simplify things, only dichotomous manifest variables will be considered in most cases in this chapter. The generalization to polytomous manifest variables is, however, straightforward.

ROLF LANGEHEINE • Institute for Science Education (IPN), University of Kiel, D-2300 Kiel 1, Federal Republic of Germany.

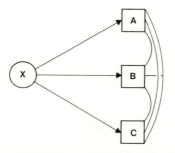

FIGURE 1. Relationships among three manifest variables A, B, and C and a latent variable X.

between them. Now consider a new categorical latent variable X so that A, B, and C are unrelated at each level of X. Expressed differently: X explains the relationships between the manifest variables. There are just three assumptions:

1. *Class Size.* The population (from which the sample was taken) consists of m latent classes of unknown size w_j, $j = 1, m$. $\sum_j w_j = 1$ because the classes are postulated to be mutually exclusive and exhaustive.

2. *Class-Specific Item Probabilities.* Within each class j, each item (indicator, manifest variable) i has a specific probability of occurrence for each of its possible outcomes c: p_{ji}^c (where $c = 1$ may denote a correct or "yes" response on a test or attitude item and $c = 0$ may denote an incorrect or "no" response). Because these p's have to obey the laws of probability, each p is a function of the other (e.g., $p_{ji}^0 = 1 - p_{ji}^1$).

3. *Local Independence.* Within each class j the manifest variables are postulated to be independent. This is the essential assumption for building homogenous groups, classes, clusters, or whatever we call them (for details, see Clogg, Chapter 8, this volume).

There are thus two kinds of parameters to be estimated: (a) The w's, that is, unconditional probabilities of being in latent class j; (b) the p's, that is, conditional probabilities of making a particular response on an item i, given membership in a certain latent class j.

This concept was developed by Lazarsfeld (1950) and elaborated on by Lazarsfeld and others. For many years, however, severe problems made people hesitate to apply LCA in substantive research:

1. Different methods of parameter estimation (Anderson, 1954; Gibson, 1955, 1962; Lazarsfeld & Dudman, 1951; Madansky, 1960) often resulted in nonadmissible estimates, that is, probabilities outside the 0–1 range.

2. The so-called basic solution (Anderson–Lazarsfeld–Dudman method) often produced different results, depending on which manifest variable was chosen as a "stratifier."

3. These methods, as well as the ML method proposed by McHugh (1956, 1958), could handle dichotomous manifest variables only.

4. All the methods were capable of handling certain specific cases of the general LCM only. It was impossible to constrain the solution according to some *a priori* hypothesis, that is, it was impossible to fix certain parameters to prespecified values or to constrain two or more parameters to be equal.

5. Only a single latent variable could be specified.

In principle, all of these problems are related to parameter estimation. The battle was won, however, when Goodman (1974a,b) showed that the parameters can in fact be estimated according to the ML method (using a special algorithm). Not only did he show how to estimate parameters in case of dichotomous or polytomous manifest variables, but he also investigated constrained LCMs as well as models with more than one latent variable.

Ever since then, LCA has been applied in various disciplines. In the following, however, I will focus on social science applications. Of course, the space available does not allow consideration of all developments published so far. My selection will thus be a subjective one, implying no devaluation of research I do not cover. Readers interested in previously published reviews are referred to Andersen (1982), Bergan (1983), Clogg (1981a), Clogg & Sawyer (1981), and Langeheine (1984).

Finally, although I had preferred to use Goodman's notational system just as many others have, I will adhere to Formann's notation in this chapter. In fact, this chapter draws heavily on Formann's (1984) book, a condensed version of which has been published only recently (1985). As Formann shows, Goodman's different LCMs are special cases[2] of what Formann calls "LCA with linearly decomposed latent parameters," and the latter stimulated the development of his even more promising linear logistic LCA. In addition, Formann notes that there is an even more general formulation, that is, Haberman's (1979) general LCM. We thus have the following hierarchy of formulations of LCMs:

1. Unconstrained LCA
2. Constrained LCA:
   - Goodman-type constraints
   - LCA with linear constraints on the parameters (Formann)

[2] In order to avoid misunderstandings: If *A* is said to be a special case of *B*, then *B* can handle all the types of LCMs that *A* can, and more.

3. Linear logistic LCA (Formann)
4. Haberman's general LCM

This hierarchy will be the main idea throughout the chapter.

2. UNCONSTRAINED LCA

There is nothing new beyond what has already been said in the introductory section. We have to estimate parameters w_j (the probability that a person belongs to class j) and p_{ji}^c (the conditional probability that a person belongs to category $c = 1$ or 0 of item i, given class j).

Data presented in Table 1 will be used for illustrative purposes in different sections of this chapter.

TABLE 1
Observed Frequencies Associated with Response Patterns of Three Test Items B, F, and D, the Two-Class Solution, and Expected Frequencies

Section I. Observed frequencies								
Item	Response patterns							
D	1	1	1	1	0	0	0	0
F	1	1	0	0	1	1	0	0
B	1	0	1	0	1	0	1	0
Observed frequency	334	53	34	63	12	43	15	501

Section II. Parameters (two-class solution)

		Conditional p's					
		B		F		D	
Class	w_j	1	0	1	0	1	0
1	.41	.87	.13	.91	.09	.97	.03
2	.59	.03	.97	.08	.92	.10	.90

Section III. Expected frequencies (two-class solution)

Class	Frequencies							
1	333.9	48.2	32.4	4.7	10.9	1.6	1.0	.2
2	.1	4.8	1.6	58.3	1.1	41.4	14.0	500.8
1 + 2	334	53	34	63	12	43	15	501

Section I of Table 1 gives frequencies observed with the respective response patterns of three dichotomous manifest variables B, F, and D.[3] Section II of Table 1 contains the parameters for the unconstrained two-class solution. We thus see that 41% of the respondents belong to Class 1. The conditional p's show that this class is characterized by a high probability of solving all three items correctly. This class may be termed a mastery class, therefore. The second class, on the other hand, may be termed a nonmastery class, because the probability of a correct response is low with all three items.

How can we evaluate whether this model fits the data? Obviously, the probability of being a member of a certain cell of the joint $B \times F \times D \times X$ table is

$$p_{s,j} = w_j \prod_i p_{ji}{}^c (1 - p_{ji})^{1-c} \tag{1}$$

where s refers to a specific response pattern. If we multiply $p_{s,j}$ by the sample size N we get the expected frequency of s in class j. Summing over j then gives the expected frequency for each response pattern (cf. Section III of Table 1). We thus see that the observed frequencies are exactly reproduced by the expected frequencies. And this is what we would expect. There are $2^3 - 1 = 7$ independent entries in the $B \times F \times D$ table and we have estimated seven nonredundant parameters (one w and six p's). The model is thus not testable, because df (degrees of freedom) $= 7 - 7 = 0$.

We may, however, ask whether there are more parsimonous two-class models (models having fewer parameters) that are congruent with the data. In fact the results of our first analysis suggest putting certain constraints on the p's.

3. CONSTRAINED LCA

3.1. GOODMAN-TYPE CONSTRAINTS

Goodman (1974a,b) considered two kinds of constraints: (a) Parameters may be fixed to prespecified values; (b) two or more parameters may be constrained to be equal.

The result of such constraints is not only that the number of parameters to be estimated may be reduced drastically, but also—and

---

[3] Though these data originate from an attitude survey (cf. Haberman, 1979), we will treat them as being test items for illustrative purposes. A 1 thus indicates that an item has been solved correctly, whereas a 0 says that a person failed to solve an item.

this is more important—that properly specified constraints enable the researcher to test a multitude of hypotheses about the structure of the data.

I will first use the three-item example and demonstrate a few possibilities of imposing constraints on the two-class solution of the preceding section and then review in more detail areas of application where constraints imply some structural hypothesis.

3.1.1. Some a Posteriori Constraints

A typical example of equality constraints imposed on the p's of the two-class solution may be the following: $p_{1B}^{1} = p_{1F}^{1} = p_{1D}^{1}$. That is, the probability of a correct response is the same for all items of the mastery class. The resulting likelihood ratio test statistic[4] $L^2 = 23.3$ (associated with 2 df) says, however, that we have to reject this hypothesis.

A second example may demonstrate the possibility of constraining parameters to fixed values: $p_{1D}^{1} = p_{2D}^{1} = 1$. This hypothesis says that D is an exact indicator of X (cf. Goodman, 1974a). X and D are therefore required to be perfectly associated, or to put it differently: X is equal to D. We no longer consider a latent variable, but test the hypothesis that B and F are independent, given D. This hypothesis does not fit either ($L^2 = 122.4$, df = 2).

3.1.2. Hypothesis Testing by a Priori Constraints

3.1.2.1. Categorical Data Analogs of Factor Analysis and Linear Structural Relation Models. The left-hand side of Figure 2, Section I, contains a path diagram corresponding to the unconstrained two-class model considered in Section II. The analogy with the one-factor model is obvious. Note, however, that there is at least one fundamental difference between LCA and classical factor analysis. Whereas factor analysis takes only first-order associations between manifest variables into account, first- as well as higher-order interactions enter into LCA. The right-hand part of Section I presents a LCM with two latent variables Y and Z. As Goodman (1974a) showed, the two latent dichotomies Y and Z may be derived from a single latent variable X having four classes. The double-headed arrow between Y and Z says that the latent variables are assumed to be associated. The single-headed arrows, on the other hand,

---

[4] Cf. Goodman (1974a,b) and others for the (dis)advantages of the likelihood ratio and Pearson chi-square statistics in evaluating model fit.

Section I: Antecedent latent variables

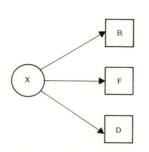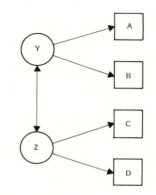

Section II: Intervening latent variables

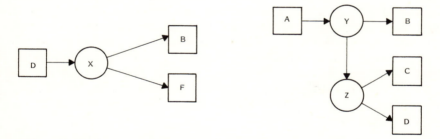

FIGURE 2. Path diagrams on the relationship of one or two latent variables with a set of manifest variables.

correspond to the hypothesis that Y affects A and B (but not C and D) and that Z affects C and D (but not A and B). The reader is referred to Goodman (1974a) for equality constraints necessary to test this hypothesis. Again, the analogy with the two-factor model with correlated factors is obvious.[5]

   Whereas all latent variables in Section I of Figure 2 are conceived as being antecedent to the manifest variables, both of the path diagrams in

---

[5] None of the LCM formulations considered in this chapter are capable of testing for orthogonal latent variables. This model, however, is equivalent to independence of (A, B) and (C, D) and may therefore be tested by fitting the corresponding log-linear model.

Section II reveal that latent variables may as well be considered as being intervening. The left-hand diagram thus specifies D to be an observed cause of X whereas B and F are considered as multiple indicators of X. Similar comments apply to the right-hand path diagram. Note, however, that both parts of Sections I and II present two different causal systems that would be analyzed by the same latent class model (cf. Goodman, 1974a, for details, as well as the possibility of assigning path coefficients to the arrows).

Because latent variables in LCA thus may be conceived as being intervening, or antecedent as well as intervening, manifold parallels of linear structural relation models for continuous data may be considered. Clogg (1981a) discusses one special case from among the family of such models: a MIMC (multiple indicator–multiple cause) model, the corresponding path diagram of which is given in Figure 3.

Note again that the latent class model corresponding to this MIMC model is identical in various aspects to the respective latent class model of the right-hand path diagram in Figure 2, Section I, despite the differences between the two causal systems. This MIMC model, however, says that two additional effects are of relevance here: C and D are associated, and, in addition to the direct effects of both C and D on Z, there is an interaction effect of C and D on Z. The interesting feature of this model is that two of the manifest variables (C and D) allowed to be associated. Clogg's model thus proposes a partial relaxation of the local independence criterion. Further references with respect to partial relaxation of local independence are Formann (1984), Harper (1972), and Langeheine (1984). Another MIMC model has been considered by Madden and Dillon (1982).

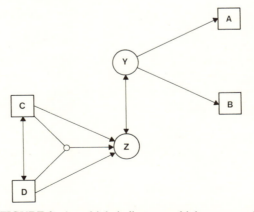

FIGURE 3. A multiple indicator–multiple cause model.

3.1.2.2. Scaling Models. Over the past 30 years many scaling models have been proposed, all of which are probabilistic generalizations of the deterministic Guttman (1950) scaling model.

For ease of exposition, let us again consider the small three-variable example from Section 2. The assumption underlying the Guttman model is that the (latent) ability is unidimensional and that the items (*B, F,* and *D*) are cumulative in their indication of different points along the (latent) scale. The first question to be answered by an analysis of Guttman scalability is the ordering of the items. Let us for simplicity use the marginal distributions of the single items in ordering them from least difficult to more difficult. This ordering is then *D, F, B* (cf. Table 2, column "Percentage correct"). A perfect Guttman scale would now request that only four response patterns (scale types, cf. Table 2) be observed with the total sample. In fact, roughly 90% of the sample belongs to any one of these scale types. Note, however, that there are at least 10% error responses.

The crucial point of the probabilistic scaling models now is the treatment of these errors. In principle, there are two concepts in handling errors: (a) latent distance models, and (b) Goodman scaling models. In addition, combinations of some features of both of these concepts have been proposed.

Latent Distance Models. Assumptions made with latent distance models (LDMs) are threefold: (a) In principle, the population is homogeneous, and unexpected response patterns are due to measurement error. (b) The number of classes is equal to the number of Guttman scale types and these are ordered. (c) Local independence holds within classes.

Let us briefly consider a test-theoretical formulation of the LDM (cf.

TABLE 2
Some Issues Relevant in a Guttman Scalability Analysis

Item	Response patterns								Percentage correct
D	1	1	1	1	0	0	0	0	46
F	1	1	0	0	1	1	0	0	42
B	1	0	1	0	1	0	1	0	37
Scale type	1	2		3				4	
Observed frequency	334	53	34	63	12	43	15	501	
Percentage	32	5		6				47	

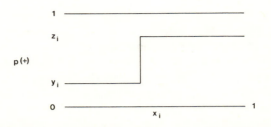

FIGURE 4. Item characteristic curve of one item in latent distance models.

Formann, 1984; Lazarsfeld & Henry, 1968). In the most general form of
the LDM, each item i is described by two parameters: y_i and z_i (cf. Figure
4). Each item i cuts the latent dimension X (and the population) into two
parts (with relative sizes x_i and $1 - x_i$). The probability of a positive
response $p(+)$ on the item i is y_i for persons below the cutting point x_i
and z_i for persons above x_i. The item characteristic curve (ICC) of each
item thus is a step function. It follows that all persons belonging to a
certain class are characterized by a specific combination of low (y_i) and
high (z_i) item probabilities. If we translate these considerations into a
LCM the parameters of the most general LDM for $k = 3$ items are those
displayed in Table 3. We will refer to this model as LDM0. Note that
with LDM0 all ICCs are asymmetric (i.e., $y_i + z_i \neq 1$) and different for
all items. With three items only, this model is not testable, because there
are more parameters to be estimated (three nonredundant w's and six
p's) than there are cells in the observed crossclassification. But even with
more than three items, this model has a fundamental indeterminacy, as
has been pointed out already by Lazarsfeld and Henry (1968).

TABLE 3
Parameters of LDM0 for Three
Dichotomous Items and Four
Classes

Class	Size	Conditional p's for item		
		1	2	3
1	w_1	y_1	y_2	y_3
2	w_2	z_1	y_2	y_3
3	w_3	z_1	z_2	y_3
4	w_4	z_1	z_2	z_3

In order to make the LDM testable, some constraints must therefore be imposed. Different versions considered are as follows:

LDM1. This model (originally proposed by Lazarsfeld & Henry, 1968) requires symmetric ICCs for the first and last item whereas the rest of the items stay with asymmetric ICCs (i.e., $y_1 = 1 - z_1$ and $y_k = 1 - z_k$). Again, ICCs differ for all items. With our three-variable example, LDM1 is not testable either, because the number of parameters estimated (7) equals the number of independent data points in the $B \times F \times D$ table.

LDM2. Both Lazarsfeld and Henry (1968) and Proctor (1970) have considered this model that imposes constraints $y_i = 1 - z_i$, so that all ICCs are symmetric but different for all items. LDM2 has also been labeled "item specific error rate model." As Table 4 shows, these error rates are estimated to be .04, .08, and .03, respectively. Whereas members of Class 1 are expected to have zero probability in solving all items in the Guttman model, at least a small proportion of members of this class passes these items under LDM2. These are the false positives, therefore. The counterpart of Class 1—Class 4—shows that not all items are solved perfectly as would be expected in the Guttman model. $y_i = 1 - z_i$ thus give the false negatives here. Because $L^2 = .39$ (df = 1), LDM2 provides a fairly good description of the data.

LDM3. Constraints assumed under LDM3 (Dayton & Macready, 1976) are $y_i = y$, $z_i = z$. All ICCs are therefore identical but asymmetric. LDM3 thus assumes a single false positive error rate (.05 here) and a

TABLE 4
Parameters for LDM2 and LDM3

Model	Class	w_j	Conditional p's (correct response) for item		
			D	F	B
LDM2	1	.54	.04	.08	.03
	2	.05	.96	.08	.03
	3	.04	.96	.92	.03
	4	.37	.96	.92	.97
LDM3	1	.55	.05	.05	.05
	2	.04	.94	.05	.05
	3	.03	.94	.94	.05
	4	.38	.94	.94	.94

single false negative error rate (.06 here; cf. Table 4). Because $L^2 = 21.66$ and df = 2, we have to reject this model.

LDM4. This is the so-called Proctor (1970) model assuming $y_i = 1 - z_i = y$. We thus have symmetric and identical ICCs for all items, that is, a single error rate. Because LDM3 does not fit, LDM4 will not fit either ($L^2 = 21.97$, df = 3).

The perfect Guttman scale, of course, would require that $y = 0$. Because both LDM3 and LDM4, which may be considered as Guttman models with weakened assumptions, do not fit, we see that the Guttman model does not fit either.

Goodman Scaling Models. Again, there are three essential assumptions in the scaling model proposed by Goodman (1975): (a) Error responses are due to heterogeneity of the population. One part of the population is scalable in the sense of Guttman, whereas a second part is not. (b) We thus have as many types/scales as in the Guttman model plus one class for the unscalables. (c) Local independence holds within the class of unscalables.

The parameters of this model (M_1) are given in Table 5. In order to

TABLE 5

Parameters for Two Goodman Scaling Models and Expected Frequencies under M_2

Model	Class	w_j	Conditional p's (correct response) for item		
			D	F	B
M_1	1	.2711	.60	.46	.30
	2	.4340	0	0	0
	3	.0004	1	0	0
	4	.0001	1	1	0
	5	.2944	1	1	1
M_2	1	.27	.60	.46	.29
	2	.44	0	0	0
	3	.29	1	1	1

Expected frequencies under M_2

Class	Frequencies							
1	23.39	27.64	15.29	18.07	56.07	66.29	36.65	43.32
2								457.68
3	310.61							
Total	334	27.64	15.29	18.07	56.07	66.29	36.65	501

assure that the "true types" (latent scalables or intrinsically scalables, as Goodman calls them) correspond to Goodman's scale types, conditional p's of Classes 2–5 have to be fixed deterministically. No constraints are made, however, for the intrinsically unscalables, that is, Class 1.

M_1 is not identified, however, and this is obviously due to the fact that Classes 3 and 4 are estimated to have negligible probability of occurrence (.0004 and .0001). We could thus fit a model without these two classes (cf. M_2, Table 5). This results in two classes corresponding to "true" masters and nonmasters, respectively, as well as a third class of unscalables, with an acceptable fit ($L^2 = 4.06$, df = 2). Table 5 also contains frequencies expected under M_2. It should be noted that two of the observed frequencies (those corresponding to response patterns 111 and 000) are fitted perfectly.[6] However, not all of the persons showing these response patterns will be grouped into Classes 2 or 3. The reason is simply that the unscalable type is assumed to generate *all* 2^k response patterns. In order for independence to hold in Class 1, we thus have to borrow from Classes 2 and 3. As Clogg and Sawyer (1981) pointed out, such a model has no counterpart in other models for item analysis. M_2 is but one possible restriction of the general model M_1. Other restrictions may be possible by constraining conditional p's of the unscalables to be equal. On the other hand, it may be possible to extend M_1 by postulating more than $k + 1$ scalable classes. Goodman (1975) considers a variety of such models. The logic behind these models is that the order of the items along the latent continuum must not necessarily be the same for all persons.

Combinations of LDMs and Goodman Scaling Models. Dayton and Macready (1980) have criticized Goodman's model for several reasons. They proposed combining certain features of LDMs with those of Goodman's model. Their models thus maintain the assumption that one part of the population is scalable whereas another part is not (Goodman model). However, conditional p's for the scalables are no longer 1 or 0 but are assumed to obey to one of the LDMs.

Many of these and other scaling models have been used in the assessment of mastery and hierarchical ordering of intellectual skills. Some references are Bergan (1983), Bergan and Stone (1985), Haertel (1984a,b), Macready and Dayton (1977, 1980), and Rindskopf (1983).

3.1.2.3. Latent Agreement Models. As Bergan (1983, p. 324) notes, the "assessment of observer agreement in the behavioral sciences has

---

[6] Both M_1 and M_2 are also called quasi-independence models because they are equivalent to quasi-independence log-linear models blanking out, that is, fitting perfectly, the cells corresponding to the "true" types.

been plagued by several difficulties." Bergan himself, as well as Clogg (1979, 1981a), has demonstrated the advantages of LCMs in this context: LMCs (a) take account of chance agreement, (b) afford coefficients with a directly interpretable meaning, and (c) identify significant sources of (dis)agreement.

A more recent paper dealing with this topic is the one by Dillon and Mulani (1984). Their data are given in Table 6. Three judges (J_1, J_2, J_3)

TABLE 6
Dillon and Mulani (1984) Data and Parameters for the
Unconstrained Three-Class Model (cf. Text)

		Data		
			Judge 3	
Judge 1	Judge 2	Positive	Neutral	Negative
Positive	positive	56	12	1
	neutral	1	2	1
	negative	0	1	0
Neutral	positive	5	14	2
	neutral	3	20	1
	negative	0	4	7
Negative	positive	0	0	2
	neutral	0	4	1
	negative	1	2	24

Parameters for unconstrained three-class model

		Agreement classes		
Judge	Category[a]	Positive (1)	Neutral (2)	Negative (3)
1	1	.83	.12	.05
	2	.15	.79	.06
	3	.02	.09	.89
2	1	.98	.33	.08
	2	.01	.55	.04
	3	.01	.12	.88
3	1	.93	.14	.01
	2	.06	.75	.21
	3	.01	.11	.78
	w_j	.41	.36	.23

[a] 1 = positive; 2 = neutral; 3 = negative.

categorized 164 respondents as being either positive, neutral, or negative toward some object. We thus start from a $3 \times 3 \times 3$ table of counts, some cells of which reflect perfect agreement among judges (e.g., Cell 111), whereas other cells reveal partial agreement (e.g., Cell 112) or no agreement at all (e.g., Cell 123).

We are told the whole story about these data if we fit an unconstrained model having three so-called agreement classes. Parameters of this well-fitting model[7] ($L^2 = 5.73$, df = 6) are summarized in Table 6 as well. If agreement among judges is good, we would expect high p's in the diagonal cells, as is in fact true. Note that Judge 2, for example, does best in the positive category, whereas his errors turn out to be relatively high in the neutral category. Apart from looking at single p's, we may also be interested in comparing the overall latent reliability of judges. This may be done by computing J's estimated overall error rate, which is equal to the weighted sum of his individual errors. For J1, for example, this error rate turns out to be $.41(.15 + .02) + .36(.12 + .09) + .23(.05 + .06) = .17$. In most cases, researchers are also interested in the reliability of assignment categories. Respective answers are given by the current approach. The probability of (for example) a neutral error in the positive agreement class is $[.41(.15 + .01 + .06)]/3 = .03$.

Because the unconstrained model fits, it is possible, of course, to look for more restrictive models by imposing one set of constraints or the other. Dillon and Mulani considered at least six such models.

3.1.2.4. Simultaneous LCA for Several Groups. Suppose we have obtained data from several groups pertaining to the same set of manifest variables. The grouping variable may be, for example, sex or different points in time. What people did for some time when analyzing such data was to perform separate LCAs for each group, followed by a more informal comparison of the results. The situation changed dramatically in 1981, when the first of a series of papers by Clogg and Clogg and Goodman were circulated, although not published until 1984–1985. These authors show how to perform a simultaneous analysis across several groups by using a very simple trick. All one needs to do is to add the group variable to the table of observed counts and to fix the conditional p's of this variable deterministically.

Table 7 presents results of such an analysis using the $B \times F \times D$ example, but for $G = 2$ groups. The hypothesis tested here is that two

---

[7] Because Dillon and Mulani were interested in standard errors of the parameters, a constant of 0.5 was added to all frequencies.

TABLE 7
Simultaneous LCA for $G = 2$ Groups
($B \times F \times D$ Data; cf. Text)

Class	w_j	B	F	D	G
1	.20	.90	.93	.97	1
2	.30	.03	.07	.10	1
3	.24	.90	.93	.97	0
4	.26	.03	.07	.10	0

classes each are adequate in mapping the data and that the two groups are homogeneous with respect to the item structure. We thus have to constrain the respective conditional p's to be equal for Groups 1 and 2. The fit of this model turns out to be extremely good: $L^2 = 7.58$, df = 6.

This across-group procedure may be applied to all variants of LCMs considered so far. But the combination of homogeneity constraints across groups (none, some, or all) with constraints within groups (none, some, or all) offers a wide field for testing new hypotheses. To give but two examples: it may be assumed that groups are equal in class proportions or that groups differ with respect to number of classes. Some references of applied research are Bergan (1983), Clogg (1984), Clogg and Goodman (1984, 1985), Dillon, Madden, and Kumar (1983), Langeheine (1984), and Tuch (1984).

A somewhat different procedure for across-group comparisons has been proposed by Macready and Dayton (1980) in the context of testing state mastery models. Instead of a group variable, these authors introduce a covariate. If the conditional p's of the covariate are fixed deterministically, this procedure is equivalent to the Clogg and Goodman approach (see also Dayton & Macready, Chapter 6, this volume). In the more general $\alpha\beta\delta$ model, however, the covariate has the status of an additional manifest variable and the respective conditional p's (the δ's) are free to vary. Their model thus allows for differential error rates for subjects belonging with a specified level of the covariate and a given state of mastery. Though this model has been presented in a specific context, generalizations to other contexts are obvious.

3.2. LCA with Linear Constraints on the Parameters

As Formann (1984, 1985) shows, Goodman-type constraints are special cases of what Formann calls "LCA with linearly decomposed latent parameters." The main feature of this approach is that the original parameters of the unconstrained LCA are related to some so-called basic parameters by two systems of linear equations. The model equations are

$$p_{ji} = \sum_{r}^{t} q_{jir}\lambda_r + c_{ji} \tag{2}$$

and

$$w_j = \sum_{s}^{u} v_{js}\eta_s + d_j \tag{3}$$

where indices $j = 1, m$ refer to classes, $i = 1, k$ to items, $r = 1, t$ to basic parameters λ ($t \leq mk$), and $s = 1, u$ to basic parameters η ($u \leq m - 1$). The weights q_{jir} and v_{js} are known constants. They result from the structural hypothesis to be tested. The additive terms c_{ji} and d_j are also known constants. They allow conditional p's and/or class proportions to be fixed at prespecified values. The λ's and η's are the unknown parameters to be estimated. The advantage of this approach as compared with Goodman-type constraints is that more general hypotheses may be formulated, for example, (a) a certain p_{ji} may be constrained to be twice as large as another one; (b) a certain w_j should be equal to the sum of two other w's.

Some examples may help to understand this procedure. Section I of Table 8 gives the class and item structure for the unconstrained two-class model considered in Section 2. Section II of Table 8 corresponds to the model constraining all response probabilities of the mastery class to be equal, and Section III contains the model in which D is considered to be an exact indicator of X (cf. Section 3.1.1). In Section IV no constraints are imposed on the response probabilities of the mastery class, but in the nonmastery class the probability of a correct response on item F is required to be twice as large as that of item B, and that of item D is requested to be three times as large as that of item B. In Section V we have the same constraints for the nonmastery class as in Section IV and, in addition, we constrain conditional p's (incorrect response) of the mastery class to be a multiple of item D, i.e., $p_{1F}{}^0 = 3p_{1D}{}^0$ and $p_{1B}{}^0 = 4p_{1D}{}^0$.

It should be noted, however, that some of the above-mentioned extended types of constraints may be easily incorporated into a Goodman-type LCA program such as MLLSA (Clogg, 1977). Because Formann's linear logistic LCA is capable of handling even more extended hypotheses, I will not go into further details.

4. LINEAR LOGISTIC LCA

The model underlying linear logistic LCA has been presented in a series of papers by Formann (1982, 1984, 1985). The basic motivation

TABLE 8
Examples of Class and Item Structures in Formann's LCA with Linear
Constraints (cf. Text)

Section	Class	Class structure Matrix V η	Vector d	Item	Item structure Matrix Q λ_1	λ_2	λ_3	λ_4	λ_5	λ_6	Vector c
I	1	1	0	D	1	0	0	0	0	0	0
				F	0	1	0	0	0	0	0
				B	0	0	1	0	0	0	0
	2	−1	1	D	0	0	0	1	0	0	0
				F	0	0	0	0	1	0	0
				B	0	0	0	0	0	1	0
II	1	1	0	D	1	0	0	0			0
				F	1	0	0	0			0
				B	1	0	0	0			0
	2	−1	1	D	0	1	0	0			0
				F	0	0	1	0			0
				B	0	0	0	1			0
III	1	1	0	D	0	0	0	0			1
				F	1	0	0	0			0
				B	0	1	0	0			0
	2	−1	1	D	0	0	0	0			0
				F	0	0	1	0			0
				B	0	0	0	1			0
IV	1	1	0	D	1	0	0	0			0
				F	0	1	0	0			0
				B	0	0	1	0			0
	2	−1	1	D	0	0	0	3			0
				F	0	0	0	2			0
				B	0	0	0	1			0
V	1	1	0	D	−1	0					1
				F	−3	0					3
				B	−4	0					4
	2	−1	1	D	0	3					0
				F	0	2					0
				B	0	1					0

behind this development was the well-known fact that older estimation methods often resulted in nonadmissible estimates of the latent probabilities outside the 0–1 range (cf. Section 1). Formann therefore proposed a parametrization using real-valued auxiliary parameters x_{ji} and z_j, where

$$x_{ji} = \log \frac{p_{ji}}{1 - p_{ji}} \tag{4}$$

and

$$z_j = \log w_j - \frac{1}{m} \sum_l^m \log w_l \tag{5}$$

with restriction $\sum_l^m z_l = 0$ (see also Formann 1976a,b; 1978a,b for this log transformation). The x's thus are log-odds, where the odds give the ratio of two probabilities, for example, to get an item right (p_{ji}) and to be wrong ($1 - p_{ji}$). The z's are log-odds of class j relative to the mean of all classes. It follows that

$$p_{ji} = \frac{\exp(x_{ji})}{1 + \exp(x_{ji})} \tag{6}$$

and

$$w_j = \frac{\exp(z_j)}{\sum_l^m \exp(z_l)} \tag{7}$$

Just as in LCA with linear constraints on the parameters (see Section 3.2), the x's and z's are expressed by weighted sums of a set of unknown "basic" parameters. The model equations thus are

$$x_{ji} = \sum_r^t q_{jir}\lambda_r + c_{ji} \tag{8}$$

and

$$z_j = \sum_s^u v_{js}\eta_s + d_j \tag{9}$$

where indices (j, i, r, s), weights, constants, and parameters correspond to those in (2) and (3). The only difference in (8) as compared with (2), and (9) as compared with (3) is that we now have the x's and z's on the left of the equal sign instead of the p's and w's. The parameters are thus first transformed using the logistic function, and linear restrictions are then imposed on these transformed parameters.

Linear logistic LCA can handle all types of LCAs considered so far, and more. Formann (1984, 1985) considers the following examples:

LCMs with Located Classes. Instead of equating each class with a certain range of the latent continuum (as in LDMs) it may be assumed that each class has a certain scale value, that is, all persons belonging to a certain class are concentrated at just this point on the scale. In order to solve this problem Lazarsfeld and Henry (1968) considered polynominal ICCs instead of step functions, however, using the older estimation methods. Formann has proposed (and tested successfully) a simple model having scale values for classes and items:

$$p_{ji} = \frac{\exp(\xi_j + \sigma_i)}{1 + \exp(\xi_j + \sigma_i)} \tag{10}$$

It is thus assumed that the ICCs follow the logistic function and that each class j and each item i is characterized by just one parameter.

As an aside, Formann stresses the analogy with Rasch's (1960) model and notes where both of these models have common features or where they differ.

Item Difficulty and Item Structure. An extension of (10) would be to let item parameters vary across classes:

$$p_{ji} = \frac{\exp(\xi_j + \sigma_{ji})}{1 + \exp(\xi_j + \sigma_{ji})} \tag{11}$$

where, however, the σ_{ji} are subject to certain linear restrictions. The specific feature of this model is that one may consider in more detail the effect of certain aspects of an item hypothesized to affect the difficulty of an item (see Formann, 1982, for details).

Measurement of Latent Change. Suppose we have measured two items at two points in time, that is, we have data from a typical panel study. The feature of linear logistic LCA then is that one may test hypotheses about specific change parameters. Formann demonstrates how to test such hypotheses for a single set of data, or where latent change has to be evaluated simultaneously across several groups. However, I will not go into details since—at least to my mind—these ideas are easier to understand if Haberman's (1979) general model is applied (see Section 6).

5. HABERMAN'S GENERAL LCM

"In general, any latent-class model can be described in terms of a three-dimensional array representing a cross-classification of N

TABLE 9
The Three Dimensions of Haberman's
General LCM

III	II				I	
	Response patterns of items				Classes	
Groups	B	F	D		1	2 · · · r
1	1	1	1	1		
	0	1	1	2		
	1	0	1			
	0	0	1			
	1	1	0	⋮		
	0	1	0			
	1	0	0			
	0	0	0	s = 8		
2				1		
				2		
				⋮		
				8		
⋮						
t				1		
				2		
				⋮		
				8		

polytomous variables" (Haberman, 1979, p. 561). As Table 9 shows, these dimensions are classes, response patterns (eight in the case of our $B \times F \times D$ table), and groups. The problem to be solved then is to estimate the $t \times s \times r$ expected frequencies under some specified hypothesis.

Note that the possibility of doing simultaneous LCAs across groups was thus presented as early as 1979. And Haberman even used a set of data from three groups in testing different models. But the application of Haberman's LAT program seems to be so difficult that people began to work with it only recently.

How does Haberman estimate the expected frequencies? In short, his formulation is equivalent to that of a log-linear model. Let us consider our example with three items, one group, and two classes without any constraints (cf. Section 2). Then it is assumed that

$$\log m_{ijkx} = \lambda + \lambda_x^X + \lambda_i^B + \lambda_j^F + \lambda_k^D + \lambda_{xi}^{XB} + \lambda_{xj}^{XF} + \lambda_{xk}^{XD} \quad (12)$$

with the usual restrictions on all indexed λ's (e.g. $\sum_x \lambda_x^X = 0$). The natural logarithm of the expected cell counts is therefore equal to a sum of parameters, where λ denotes the grand mean, λ_x^X represents category x of the latent variable X, λ_i^B, λ_j^F, and λ_k^D stand for the manifest variables B, F, and D (all of which have two categories), and the interaction terms λ_{xi}^{XB}, λ_{xj}^{XF}, and λ_{xk}^{XD} capture the effects of the latent variable X on the three manifest variables.

LAT now requires the user to supply a design matrix, among other things. Two examples of design matrices are given in Table 10, where the matrix containing columns 1–7 corresponds to model (12). Note the

TABLE 10
Two Examples of a LAT Design Matrix (cf. Text)

Column:	1	2	3	4	5	6	7	8
Effect:	X	B	F	D	XB	XF	XD	$XB + XF + XD$
Response pattern								
1	1	1	1	1	1	1	1	3
	−1	1	1	1	−1	−1	−1	−3
2	1	−1	1	1	−1	1	1	1
	−1	−1	1	1	1	−1	−1	−1
3	1	1	−1	1	1	−1	1	1
	−1	1	−1	1	−1	1	−1	−1
4	1	−1	−1	1	−1	−1	1	−1
	−1	−1	−1	1	1	1	−1	1
5	1	1	1	−1	1	1	−1	1
	−1	1	1	−1	−1	−1	1	−1
6	1	−1	1	−1	−1	1	−1	−1
	−1	−1	1	−1	1	−1	1	1
7	1	1	−1	−1	1	−1	−1	−1
	−1	1	−1	−1	−1	1	1	1
8	1	−1	−1	−1	−1	−1	−1	−3
	−1	−1	−1	−1	1	1	1	3

following advantages of the design matrix approach:

1. The user is free to choose from among different ways of coding.

2. LAT allows certain constrained hypotheses to be tested which may not be evaluated by Goodman-type constraints. The results of the saturated model (12) with respect to the interaction terms are that the three λ's are fairly equal ($\lambda^{XB} = 1.38$, $\lambda^{XF} = 1.21$, $\lambda^{XD} = 1.39$). The hypothesis that $\lambda^{XB} = \lambda^{XF} = \lambda^{XD}$, which is Haberman's model of constant interaction, can thus be tested by simply adding the respective columns of the design matrix. The design matrix now contains Columns 1–4 and 8 (see Rindskopf, 1984, for this as well as other types of constraints). And this model fits well ($L^2 = 4.06$, df = 2).

3. The Haberman formulation is even more general than Formann's linear logistic LCA, as it allows (in Formann's terminology) for different weights depending on the class and response pattern considered.

The main disadvantage with the design matrix approach, however, is that the user may be required to generate a huge matrix. Unfortunately, there are other problems involved with LAT. The user has to supply initial estimates of either the parameters or the expected counts. The trouble now is that these have to be relatively close to the final estimates. Otherwise the program may descend. A situation could thus occur where it is not known whether the design matrix or the initial estimates (or even both) are at fault.

In closing this section, I would like to stress another advantage of LAT as compared to the most widely used MLLSA program (Clogg, 1977). LAT is more rigorous in testing for local identifiability of the model parameters. There may be situations in which results from MLLSA reveal certain peculiarities, though the identification test signals okay. On the other hand, there may be situations where one is fooled into assuming that everything is okay, though in fact it is not. LAT will tell exactly what is wrong in such cases.

6. ANALYSIS OF SEQUENTIAL CATEGORICAL DATA

The space available here does not allow consideration of all types of problems tackled so far in the relevant literature. I will therefore sketch briefly what might be subsumed under this heading.

In general, sequential categorical data may be of the following types:

1. One variable is measured at two points in time, but *not* with the same persons at t_1, t_2. Some typical examples are the following:

Mobility Table Analysis. For more than two decades, sociologists have been interested in so-called mobility tables, that is, two-way tables where the two manifest variables are occupational status of father and son (often denoted by O = origin and D = destination; cf. Figure 5). Clogg (1981b) has demonstrated the utility of LCA in this context by showing how different hypotheses may be interpreted as latent structures, constrained latent structures, or quasi-latent structures.

The common feature of all these models is that a latent variable X, which can explain the observed relationships between father's and son's occupational statuses, is postulated to intervene between O and D, where the number of classes of X depends on the substantive hypothesis to be tested. The path diagram on the right of Figure 5, e.g., corresponds to a constrained two-class model saying (a) that two classes (which may be termed "upper" and "lower" class) might suffice in explaining the data, and, (b) that mobility from O to D is allowed for only within classes with the exception of status Category 3.

Party Preferences of Parents and Their Offspring. Models similar to those of Clogg are investigated by Dillon *et al.* (1983) in the context of parent–child party preferences.

2. One variable is measured at several points in time, but with the *same persons* at t_1, t_2 $(t_3 \cdots)$. Some substantive issues considered with this type of data are as follows:

Response Consistency over Time. LCMs considered to assess response consistency follow either the tradition of the Goodman scaling model (cf. Section 3.1.2.2, i.e., consistent classes correspond to "true" types or scalables and inconsistent responses are recruited into the class of unscalables) or allow for errors in latent classes reflecting response consistency. Models proposed by both Bergan (1983) and Clogg (1984) belong to the first category. Whereas Bergan considers an extremely restricted three-class model for a 2×2 table only, Clogg uses the

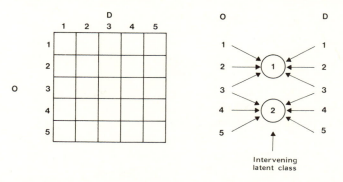

FIGURE 5. Two-way mobility table and path diagram representing a constrained two-class structure. Manifest variables: O = origin status; D = destination status.

across-group approach in order to examine the effects of various constraints in case of two 3×3 test–retest tables where the group variable corresponds to the retest condition in two consecutive years. Models of this type have also been used in Taylor (1983) in testing the so-called black-and-white model of attitude stability for a set of panel data, or may be used in testing the mover–stayer model of Blumen, Kogan, and McCarthy (1955). All models proposed by Dayton and Macready (1983), on the other hand, postulate a certain number of consistent classes only (where the number of classes depends on the data examined, i.e., dichotomous or polytomous, repeated classifications or repeated multiple classifications) and account for response inconsistency by different types of error rates.

Markov Chain Models. Wiggins (1955, 1973; cf. also Lazarsfeld & Henry, 1968) has proposed various models focusing on change from one point of time to another. However, his results have been obtained with one of the older, insufficient methods of parameter estimation. The work of Poulsen (1982), who presents reformulations of these models using the algorithm considered by Goodman (1974a), may therefore be greatly appreciated. Among other things, the two most prominent models considered are the mixed Markov model (MM) and the latent Markov model (LM). Of course, data of this kind may be analyzed by a traditional LCM, and the conditional p's would tell us something about the developmental process within classes. In this context, however, the LCM has the disadvantage that it does not allow for feedback effects from $t - 1$ on t that are quite likely to be relevant. Both the MM and the LM model do take such effects into consideration, though in a different way.

The parameters of the first-order MM model are (a) class-proportions, (b) conditional p's at t_1 (given class j), and (c) transition probability matrices denoting the conditional probability at time $t + 1$ given time t (and given class j).

If all transition probability matrices are constrained to be equal (within classes) this is a stationary first-order MM. Results from Poulsen (1982), who analyzed data from a sequence of five waves with respect to purchase of some brand A versus $O =$ all other brands, may help to better understand the MM. Table 11 reproduces results of the stationary first-order two-class MM favored by Poulsen. Class 2, for example, contains 81% of the consumers, and the odds of buying O versus A are .87:.13 or about 7:1. The process thus starts with a high probability for choosing O, whereas in Class 1 the odds are about 3:1 in favor of A. Both classes reveal different switching patterns. Class 2 has a very high repeat probability in case of O (.94) and a low repeat probability in case

TABLE 11
Parameters of the MM and LM Models

Model MM				
	Class 1		Class 2	
Class proportions	.19		.81	
	O	A	O	A
Conditional p's at t_1	.32	.68	.87	.13
	O	A	O	A
Stationary transition matrices	$O\begin{bmatrix}.50 \\ .29\end{bmatrix}$ A	$\begin{bmatrix}.50 \\ .71\end{bmatrix}$	$\begin{bmatrix}.94 \\ .79\end{bmatrix}$	$\begin{bmatrix}.06 \\ .21\end{bmatrix}$

Model LM		
	Class 1	Class 2
Initial class proportions	.25	.75
Conditional p's in favor of A	.77	.05
Stationary transition matrix	$\begin{bmatrix}.81 \\ .02\end{bmatrix}$	$\begin{bmatrix}.19 \\ .98\end{bmatrix}$ Class 1 / Class 2

of A (.21). The odds in favor of switching back to O from A thus are .79:.21 or about 4:1. Though the MM model may have considerable appeal in many other contexts, it should be noted that it is based on a simplifying assumption with respect to the underlying stochastic process: class membership remains constant across points in time. This assumption is relaxed in the LM model where individuals are allowed to change latent positions (or to make "real" change).

The parameters of the stationary LM model thus are (a) initial class proportions, (b) conditional p's, and (c) stationary transition probabilities denoting probabilities of staying with classes or switching from one class to another.

The results in Table 11 with respect to the LM model may be summarized as follows. The process starting at t_1 depicts 75% of the consumers as belonging to Class 2, which is characterized by a very low probability of choosing brand A. The probability of staying in this class is .98. Because the probability of switching into this class is about ten times as large as the probability of leaving this class (.19 vs. .02), the size of Class 2 will increase in the long run—that is, the dynamics of the market are adverse to brand A.

Note that Bye and Schechter (1986) presented a similar solution to the latent Markov chain model.

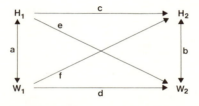

FIGURE 6. Husband's and wife's behavior at two points in time. H_1 = husband's behavior at time t_1; H_2 = husband's behavior at t_2; W_1 = wife's behavior at t_1; W_2 = wife's behavior at t_2; a, b = synchronous relationships; c, d = autodependence; e, f = dominance or cross-lagged dependence.

3. One variable is measured at two points in time, but with two sets of persons: Several hypotheses for typical data of this type have been considered by Dillon *et al.* (1983). Figure 6 depicts some relationships that may be of interest in analyzing husband's and wife's behavior at t_1 and t_2, where, among other things, Dillon *et al.* comment on the following relationships: (a) Suppose one wishes to test for dependency of H_2 on W_1. Then it is obvious that a significant relationship of this kind may be simply due to the fact that both H_2 and W_1 depend on H_1. The problem of autodependence likely to be present here may be controlled for by a constrained LCM postulating that path f is equal for both categories of H_1. (b) If, on the other hand, the researcher is interested in dominance relationships (i.e., path $f = W_1$ over H_2, path $e = H_1$ over W_2) relative to each other, any autodependence has to be partialed out. Again, a constrained four-class model may be used to test for the hypothesis that path e = path f.

4. Two variables are measured at two points in time (same persons): The Coleman panel data (cf. Goodman, 1974a) are a typical example. As Goodman showed, a constrained four-class model (with two latent variables Y and Z) corresponding to Figure 7 fits these data extremely well ($L^2 = 1.27$, df = 4). If we translate this model into the Haberman formulation, the parameters of the log-linear LCM are

$$H_0: \quad Y, Z, YZ, A, C, B, D, YA, YC, ZB, ZD$$

This model allows for change from t_1 to t_2 for both self-perception and attitude. How could we test the hypothesis of no change from t_1 to t_2? Obviously, all we have to do is to constrain two times two rows of parameters to be equal (i.e., the four p's pertaining to C should be equal to those associated with A, and row D = row B). As the corresponding

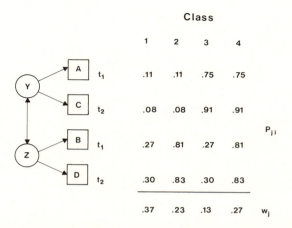

FIGURE 7. Coleman panel data. Path diagram and parameters associated with H_0 for boys. Manifest variables: A, C = self-perception of membership in leading crowd at t_1, t_2; B, D = attitude concerning leading crowd at t_1, t_2.

Haberman model

$$H_4: \quad Y, Z, YZ, A = C, B = D, YA = YC, ZB = ZD$$

shows, we require the marginal distributions of A and C (as well as B and D) to be equal and we postulate that the effect of the latent variables on the manifest variables is equal at both points in time. We thus have to estimate four parameters less. But $L^2 = 40.72$ and df $= 8$ reveal that the no-change model has to be rejected.

   The specific feature of Formann's (1984, 1985) work now is that he considers three additional models of change that are not testable by Goodman-type constraints, though they are testable by using linear logistic LCA or Haberman's general model.

   Hypothesis H_2 allows for item-specific change from t_1 to t_2. The corresponding LAT formulation is

$$H_2: \quad Y, Z, YZ, A, C, B, D, YA = YC, ZB = ZD$$

Note that this model is equal to the no-change model H_4, with the exception that A and C (and B and D, respectively) are not required to be equal. This model turns out to be acceptable ($L^2 = 5.33$, df $= 6$). We may therefore try one additional constraint, resulting in a model that postulates equal change in both items

$$H_3: \quad Y, Z, YZ, A = C, B = D, YA = YC, ZB = ZD, C = D$$

H_3, which fits as well ($L^2 = 9.92$, df $= 7$), is equal to the no-change model H_4 except for the additional parameter $C = D$. Finally,

$$H_1: \quad Y, Z, YZ, A = C, B = D, YA = YC, ZB = ZD, L_1, L_2$$

allows for class-specific change (independent of items).[8] Again, the specification is identical with that of H_4 except for L_1 and L_2, which are defined as follows for each response pattern:

	\multicolumn{4}{c}{Class}		
	1	2	3	4
L_1	$C = D$	C	D	0
L_2	0	D	C	$C = D$

This model thus says: (a) change from A to C in Classes 1 and 2, and from B to D in Classes 1 and 3 is equal to L_1; (b) change from A to C in Classes 3 and 4, and from B to D in Classes 2 and 4 is equal to L_2. Because this model fits as well ($L^2 = 5.01$, df $= 6$), we have at least three reasonable explanations of change captured by H_0. Of course, other models of change might be considered (see e.g., Andersen, Chapter 9, this volume, for a comparison of fitting latent trait and latent class models to these data), but things become extremely interesting only when Formann performs a simultaneous analysis for boys and girls, one result of which is that is now possible to decide between rival hypotheses accepted for the boys-only sample.

7. REFERENCES

Andersen, E. B. (1982). Latent structure analysis: A survey. *Scandinavian Journal of Statistics, 9*, 1–12.

Anderson, T. W. (1954). On estimation of parameters in latent structure analysis. *Psychometrika, 19*, 1–10.

Bergan, J. R. (1983). Latent-class models in educational research. In E. W. Gordon (Ed.), *Review of research in education 10*. Washington, D.C.: American Educational Research Association.

[8] Though this is the only model resulting in class proportions differing from those of all other models, it should be noted that H_1 is no "real" change model in the sense of the LM model considered earlier.

Bergan, J. R., & Stone, C. A. (1985). Latent class models for knowledge domains. *Psychological Bulletin, 98,* 166–184.

Blumen, I. M., Kogan, M., & McCarthy, P. J. (1955). *The industrial mobility of labor as a probability process.* Ithaca: Cornell University Press.

Bye, B. V., & Schechter, E. S. (1986). A latent Markov model approach to the estimation of response errors in multivariate panel data. *Journal of the American Statistical Association, 81,* 375–380.

Clogg, C. C. (1977). *Unrestricted and restricted maximum likelihood latent structure analysis: A manual for users* (Working Paper 1977-09). University Park: Population Issues Research Center.

Clogg, C. C. (1979). Some latent structure models for the analysis of Likert-type data. *Social Science Research, 8,* 287–301.

Clogg, C. C. (1981a). New developments in latent structure analysis. In D. J. Jackson & E. F. Borgatta (Eds.), *Factor analysis and measurement in sociological research.* London: Sage.

Clogg, C. C. (1981b). Latent structure models of mobility. *American Journal of Sociology, 86,* 836–868.

Clogg, C. C. (1984). Some statistical models for analyzing why surveys disagree. In C. F. Turner & E. Martin (Eds.), *Surveying subjective phenomena* (Vol. 2). New York: Russell Sage Foundation.

Clogg, C. C., & Goodman, L. A. (1984). Latent structure analysis of a set of multidimensional contingency tables. *Journal of the American Statistical Association, 79,* 762–771.

Clogg, C. C., & Goodman, L. A. (1985). Simultaneous latent structure analysis in several groups. In N. B. Tuma (Ed.), *Sociological methodology 1985.* San Francisco: Jossey-Bass.

Clogg, C. C., & Sawyer, D. O. (1981). A comparison of alternative models for analyzing the scalability of response patterns. In S. Leinhardt (Ed.), *Sociological methodology 1981.* San Francisco: Jossey-Bass.

Dayton, C. M., & Macready, G. B. (1976). A probabilistic model for validation of behavioral hierarchies. *Psychometrika, 41,* 189–204.

Dayton, C. M., & Macready, G. B. (1980). A scaling model with response errors and intrinsically unscalable respondents. *Psychometrika, 45,* 343–356.

Dayton, C. M., & Macready, G. B. (1983). Latent structure analysis of repeated classifications with dichotomous data. *British Journal of Mathematical and Statistical Psychology, 36,* 189–201.

Dillon, W. R., & Mulani, N. (1984). A probabilistic latent class model for assessing inter-judge reliability. *Multivariate Behavioral Research, 19,* 438–458.

Dillon, W. R., Madden, T. J., & Kumar, A. (1983). Analyzing sequential categorical data on dyadic interaction: A latent structure approach. *Psychological Bulletin, 94,* 564–583.

Formann, A. K. (1976a). *Schätzung der Parameter in Lazarsfeld's Latent-Class-Analysis* (Research Bulletin No. 18). Wien: Institut für Psychologie der Universität Wien.

Formann, A. K. (1976b). *Latent-Class-Analyse polychotomer Daten* (Research Bulletin No. 19). Wien: Institut für Psychologie der Universität Wien.

Formann, A. K. (1978a). A note on parameter estimation for Lazarsfeld's latent class analysis. *Psychometrika, 43,* 123–126.

Formann, A. K. (1978b). The latent class analysis of polytomous data. *Biometrical Journal, 20,* 755–771.

Formann, A. K. (1982). Linear logistic latent class analysis. *Biometrical Journal, 24,* 171–190.

Formann, A. K. (1984). *Die Latent-Class-Analyse*. Weinheim: Beltz.

Formann, A. K. (1985). Constrained latent class models: Theory and applications. *British Journal of Mathematical and Statistical Psychology, 38*, 87–111.

Gibson, W. A. (1955). An extension of Anderson's solution for the latent structure equations. *Psychometrika, 20*, 69–73.

Gibson, W. A. (1962). Extending latent class solutions to other variables. *Psychometrika, 27*, 73–81.

Goodman, L. A. (1974a). The analysis of systems of qualitative variables when some of the variables are unobservable. Part I—A modified latent structure approach. *American Journal of Sociology, 79*, 1179–1259.

Goodman, L. A. (1974b). Exploratory latent structure analysis using both identifiable and unidentifiable models. *Biometrika, 61*, 215–231.

Goodman, L. A. (1975). A new model for scaling response patterns: An application of the quasi-independence concept. *Journal of the American Statistical Association, 70*, 755–768.

Guttman, L. (1950). The basis for scalogram analysis. In S. A. Stouffer, L. Guttman, E. A. Suchman, P. F. Lazarsfeld, S. A. Star, & J. A. Clausen (Eds.), *Measurement and prediction: Studies in social psychology in World War II* (Vol. IV). Princeton: Princeton University Press.

Haberman, S. J. (1979). *Analysis of qualitative data: Vol. 2. New developments*. New York: Academic Press.

Haertel, E. (1984a). Detection of a skill dichotomy using standardized achievement test items. *Journal of Educational Measurement, 21*, 59–72.

Haertel, E. (1984b). An application of latent class models to assessment data. *Applied Psychological Measurement, 8*, 333–346.

Harper, D. (1972). Local dependence latent structure models. *Psychometrika, 37*, 53–59.

Langeheine, R. (1984). Neuere Entwicklungen in der Analyse latenter Klassen und latenter Strukturen. *Zeitschrift für Sozialpsychologie, 15*, 199–210.

Lazarsfeld, P. F. (1950). The logical and mathematical foundation of latent structure analysis. In S. A. Stouffer, L. Guttman, E. A. Suchman, P. F. Lazarsfeld, S. A. Star, & J. A. Clausen (Eds.), *Measurement and prediction: Studies in social psychology in World War II* (Vol. IV). Princeton: Princeton University Press.

Lazarsfeld, P. F., & Dudman, J. (1951). The general solution of the latent class case. In P. F. Lazarsfeld (Ed.), *The use of mathematical models in the measurement of attitudes*. Santa Monica: RAND Corporation.

Lazarsfeld, P. F., & Henry, N. W. (1968). *Latent structure analysis*. Boston: Houghton Mifflin.

Macready, G. B., & Dayton, C. M. (1977). The use of probabilistic models in the assessment of mastery. *Journal of Educational Statistics, 2*, 99–120.

Macready, G. B., & Dayton, C. M. (1980). The nature and use of state mastery models. *Applied Psychological Measurement, 4*, 493–516.

Madansky, A. (1960). Determinantal methods in latent class analysis. *Psychometrika, 25*, 183–198.

Madden, T. J., & Dillon, W. R. (1982). Causal analysis and latent class models: An application to a communication hierarchy of effects model. *Journal of Marketing Research, 19*, 472–490.

McHugh, R. B. (1956). Efficient estimation and local identification in latent class analysis. *Psychometrika, 21*, 331–347.

McHugh, R. B. (1958). Note on "Efficient estimation and local identification in latent class analysis." *Psychometrika, 23*, 273–274.

Poulsen, C. A. (1982). *Latent structure analysis with choice modeling applications*. Aarhus: Aarhus School of Business Administration and Economics.

Proctor, C. H. (1970). A probabilistic formulation and statistical analysis of Guttman scaling. *Psychometrika, 35*, 73–78.

Rasch, G. (1960). *Probabilistic models for some intelligence and attainment tests*. Copenhagen: Danish Institute for Educational Research.

Rindskopf, D. (1983). A general framework for using latent class analysis to test hierarchical and nonhierarchical learning models. *Psychometrika, 48*, 85–97.

Rindskopf, D. (1984). Linear equality restrictions in regression and loglinear models. *Psychological Bulletin, 96*, 597–603.

Taylor, M. C. (1983). The black-and-white model of attitude stability: A latent class examination of opinion and nonopinion in the American public. *American Journal of Sociology, 89*, 373–401.

Tuch, S. A. (1984). A multivariate analysis of response structure: Race attitudes, 1972–1977. *Social Science Research, 13*, 55–71.

Wiggins, L. M. (1955). *Mathematical models for the analysis of multi-wave panels*. Unpublished doctoral dissertation, Columbia University.

Wigging, L. M. (1973). *Panel analysis*. Amsterdam: Elsevier.

CHAPTER FIVE

Log-Linear Modeling, Latent Class Analysis, or Correspondence Analysis

Which Method Should Be Used for the Analysis of Categorical Data?

B. S. EVERITT AND G. DUNN

1. INTRODUCTION

Data collected by social and behavioral scientists very often consist of large multidimensional tables of subjects cross-classified according to the values or states of several categorical variables. For example, Table 1 shows a set of data on suicide victims in which the method of committing suicide is cross-classified by sex and age group (Van der Heijden & de Leeuw, 1985) and Table 2 shows counts of subjects resulting from a survey of the political attitudes of a sample from the British electorate (Butler & Stokes, 1974). The analysis of such data should clearly depend on the substantive questions posed by the researcher involved, although in many cases these questions will be rather vague. The research worker may be interested in such notions as "pattern" and "structure" but it will often be left to the statistician to clarify what is meant by such concepts and whether they are present in the investigator's data. Finally, the statistician has the often difficult task of explaining the results.

Multidimensional categorical data can be analyzed using several different methods. One of the most popular, particularly in the United

---

B. S. EVERITT AND G. DUNN • Biometrics Unit, Institute of Psychiatry, University of London, London SE5 8AF, United Kingdom.

TABLE 1
Suicide Behavior: Age by Sex by Cause of
Death[a]

Age group	Cause of death					
	M_1	M_2	M_3	M_4	M_5	M_6
Men						
10–40	3983	1218	4555	1550	550	1248
40–70	3996	826	7971	1689	517	828
70 +	938	45	3160	334	268	147
Women						
10–40	2593	153	956	141	407	383
40–70	4507	136	4528	264	715	601
70 +	1548	29	1856	77	383	106

[a] M_1 = Suicide by solid or liquid matter; M_2 = suicide by gas; M_3 = suicide by hanging, strangling, suffocating or drowning; M_4 = suicide by guns, knives, and explosives; M_5 = suicide by jumping; M_6 = suicide by other methods.

States and in Britain, involves fitting and estimating the parameters of *log-linear models* (see Everitt, 1977). An alternative, which was originally described by Green (1951), is *latent class analysis.* The third method that we shall discuss in this chapter is *correspondence analysis,* a method that is particularly popular in France, where it is often used to the complete exclusion of the others. Correspondence analysis, as it is used by the French, is particularly associated with the name of Benzécri (1969).

In this chapter the three methods mentioned above will be compared by example of their use on the data sets of Tables 1 and 2. We begin, however, with a brief description of each of log-linear modeling, latent class analysis, and correspondence analysis.

2. METHODS FOR ANALYSIS OF MULTIDIMENSIONAL CATEGORICAL DATA

2.1. LOG-LINEAR MODELING

This form of analysis originated in the work of Birch (1963), it is described in detail in the book by Bishop, Fienberg, and Holland (1975). To illustrate the approach we will consider models for a three-

TABLE 2
Voting Behavior: Vote by Sex by Class by Age[a]

Age group	Men (M)		Women (W)	
	Conservative (C)	Labour (L)	Conservative (C)	Labour (L)
Upper middle class (UMC)				
>73 (A_1)	4	0	10	0
51–73 (A_2)	27	8	26	9
41–50 (A_3)	27	4	25	9
26–40 (A_4)	17	12	28	9
<26 (A_5)	7	6	7	3
Lower middle class (LMC)				
>73 (A_1)	8	4	9	2
51–73 (A_2)	21	13	33	8
41–50 (A_3)	27	12	29	4
26–40 (A_4)	14	15	17	13
<26 (A_5)	9	9	13	7
Working class (WC)				
>73 (A_1)	8	15	17	4
51–73 (A_2)	35	62	52	53
41–50 (A_3)	29	75	32	70
26–40 (A_4)	32	66	36	67
<26 (A_5)	14	34	18	33

[a] From Payne (1977). Reprinted by permission.

dimensional table such as that given in Table 1. A log-linear model postulates that the logarithms of the expected frequencies in such a table, assuming that their sampling distribution is Poisson, can be represented as a linear function of various parameters; these parameters represent the marginal effects of each variable (for example, in Table 1 method of suicide, age group, and sex), pairs of variables, or the set of all three variables. The parameters are analogous to main effects and first- and second-order interactions in a three-way analysis of variance model. The "saturated" model for a three-variable table is given by

$$\log(m_{ijk}) = u + u_{1(i)} + u_{2(j)} + u_{3(k)} + u_{12(ij)} + u_{13(ik)} + u_{23(jk)} + u_{123(ijk)}$$

$$(1)$$

The parameters $u_{1(i)}$, $u_{2(j)}$, ..., $u_{12(ij)}$, ..., and so forth, represent the marginal effects of single variables, pairs of variables, and so forth, and are subject to constraints of the form

$$\sum_i u_{1(i)} = 0, \qquad \sum_i u_{12(ij)} = \sum_k u_{13(ik)} = 0$$

(See Bishop, Fienberg, & Holland for the explicit definition of these parameters.) The model given by (1) contains as many independent parameters as cells in the table; consequently it will fit the data exactly and the estimated expected frequencies will simply equal those observed frequencies.

The general aim in an analysis of data in terms of log-linear models will be to set some of the parameters in (1) to zero (setting those representing higher order interactions equal to zero prior to those representing low-order interactions or main effects) to obtain a hierarchical series of simpler models. The adequacy of these simpler models for the data can be tested by a comparison of the appropriate estimated expected frequencies with the corresponding observed value using a goodness-of-fit statistic such as Pearson's chi-square or a likelihood-ratio criterion.

To clarify the general approach, consider the application of this type of analysis to the data in Table 3, which has been obtained by cross-classifying subjects with respect to their hair and eye color (Fisher, 1940.) The saturated model for these data has the form

$$\log(m_{ij}) = u + u_{1(i)} + u_{2(j)} + u_{12(ij)} \qquad (2)$$

where $i = 1, 2, 3, 4$ (corresponding to light, blue, medium, or dark eyes, respectively) and $j = 1, 2, 3, 4, 5$ (corresponding to fair, red, medium, dark, or black hair, respectively.) The $u_{12(ij)}$ interaction terms are measures of association between eye and hair color, whereas the $u_{1(i)}$ and $u_{2(j)}$ terms simply reflect the marginal totals for the different eye and hair

TABLE 3
Eye Color and Hair Color Data

Eye color	Hair color				
	Fair (FH)	Red (RH)	Medium (MH)	Dark (DH)	Black (BH)
Light (Le)	688	116	584	188	4
Blue (BE)	326	38	241	110	3
Medium (ME)	343	84	909	412	26
Dark (DE)	98	48	403	681	81

colors. A simpler model for these data may be obtained by constraining all of the $u_{12(ij)}$ parameters to be zero (equivalent to the usual null hypothesis of no interaction or no association between eye and hair color.) This model is therefore, as follows:

$$\log(m_{ij}) = u + u_{1(i)} + u_{2(j)} \tag{3}$$

Fitting model (3) results in a Pearson chi-square of 1240 with 12 degrees of freedom. Clearly it is inadequate. Patterns of departure from the model can be investigated by examining the $u_{12(ij)}$ terms obtained for the saturated model, by looking at the residuals from fitting the independence model, or by examining the graphical displays produced by the use of such techniques as correspondence analysis (see later).

2.2. LATENT CLASS ANALYSIS

Latent class analysis is comprehensively described in Lazarsfeld and Henry (1968). It has its origins in sociology and psychology and is, in many respects, analogous to factor analysis of continuous measurements. The basic model postulates an underlying categorical latent variable (see Everitt, 1984), with say c classes; within any category of the latent variable the manifest or observed categorical variables are assumed independent of one another (the axiom of conditional independence). Observed relationships between the manifest variables are thus assumed to result from the underlying classification of the data produced by the categorical latent variable. The latent class model may be formulated in terms of a finite mixture density (see Everitt & Hand, 1981) as follows. Suppose $\mathbf{x}' = [x_1 \cdots x_p]$ is a vector containing the observed values of the p manifest dichotomous variables (the model is easily extended to categorical variables with more than two categories), so that x_j is either zero or one. The latent class model implies that $\mathbf{x}$ has a probability density function given by

$$f(\mathbf{x}; \alpha, \mathbf{\Theta}) = \sum_{i=1}^{c} \alpha_i f_i(\mathbf{x}; \mathbf{\theta}_i) \tag{4}$$

$$f_i(\mathbf{x}; \mathbf{\theta}_i) = \prod_{j=1}^{p} \theta_{ij}^{x_j} (1 - \theta_{ij})^{1-x_j} \tag{5}$$

and

$$\mathbf{\alpha}' = [\alpha_1 \cdots \alpha_c], \qquad \sum_{i=1}^{c} \alpha_i = 1, \qquad \mathbf{\Theta} = [\mathbf{\theta}_1 \cdots \mathbf{\theta}_c]$$

The elements of $\boldsymbol{\theta}_i' = [\theta_{i1} \cdots \theta_{ip}]$ give the probability of variables in the ith class taking the value unity. The density function, f, is a finite mixture density in which the component densities $f_1 \cdots f_c$ are multivariate Bernoulli densities arising from the local independence requirement. Formulating the latent class model in this way allows estimation of the parameters, $\boldsymbol{\alpha}$ and $\boldsymbol{\Theta}$, by maximum likelihood.

Suppose now we have a sample of n independent response vectors $\mathbf{x}_1 \cdots \mathbf{x}_n$. The likelihood function assuming the latent class model is given by

$$\mathcal{L} = \prod_{k=1}^{n} f(\mathbf{x}_k; \boldsymbol{\alpha}, \boldsymbol{\Theta}) \tag{6}$$

so that the log-likelihood, L, is

$$L = \sum_{k=1}^{n} \log_e \left[\sum_{i=1}^{c} \alpha_i f_i(\mathbf{x}_k; \boldsymbol{\theta}_i) \right] \tag{7}$$

Remembering that $\sum_{i=1}^{c} \alpha_i = 1$, we are, by differentiating L with respect to the parameters, led to the following estimation equations:

$$\hat{\alpha}_i = \frac{1}{n} \sum_{k=1}^{n} \hat{P}(i \mid \mathbf{x}_k) \tag{8}$$

$$\hat{\boldsymbol{\theta}}_i = \frac{1}{n\hat{\alpha}_i} \sum_{k=1}^{n} \mathbf{x}_k P(i \mid \mathbf{x}_k) \tag{9}$$

where $\hat{P}(i \mid \mathbf{x}_k)$ is the estimated value of the posterior probability that observation $\mathbf{x}_k$ arises from class i of the latent variable; this probability is given by

$$\hat{P}(i \mid \mathbf{x}_k) = \frac{\hat{\alpha}_i f_i(\mathbf{x}_k; \hat{\boldsymbol{\theta}}_i)}{f(\mathbf{x}_k; \hat{\boldsymbol{\alpha}}, \hat{\boldsymbol{\Theta}})} \tag{10}$$

Equations (8) and (9) do not, of course, give the parameter estimates explicitly since $\hat{P}(i \mid \mathbf{x}_k)$ involves the parameters in a complex way. However, writing the equations in this way does suggest a possible iterative scheme for their solution. Initial values of the elements of $\boldsymbol{\alpha}$ and of the $\boldsymbol{\theta}_1 \cdots \boldsymbol{\theta}_c$ are obtained in some way (perhaps simply a "guess") and used to provide initial values for the posterior probabilities in (10). These are then inserted in the left-hand side of equations (8) and (9) to give revised estimates and the procedure is repeated until some convergence

TABLE 4
Parameter Estimates for a
Latent Class Model Fitted
to the Data in Table 3[a]

	Class 1	Class 2
Proportion[b]	0.59	0.41
Pr(FH)	0.41	0.06
Pr(RH)	0.07	0.03
Pr(MH)	0.42	0.37
Pr(DH)	0.10	0.49
Pr(BH)	0.00	0.05
Pr(LE)	0.48	0.03
Pr(BE)	0.20	0.03
Pr(ME)	0.31	0.36
Pr(DE)	0.01	0.58

[a] $\chi^2 = 166.9$; df, 4. *Note.* Model
is not identified.
[b] The symbols FH, RH, and so
forth are explained in Table 3.

criterion is satisfied. The method is generally known as the EM algorithm
and is described in detail in Dempster, Laird, and Rubin (1977).

Fitting a two-class model to the eye-color–hair-color data in Table 3
results in the parameter estimates shown in Table 4. Although these
appear reasonably sensible in terms of the data, they should not be taken
too seriously because using the method described in Goodman (1974) for
studying local identifiability shows that the model is not identifiable.

2.3. CORRESPONDENCE ANALYSIS

Correspondence analysis is traditionally associated with two-way
contingency tables, and has been described by a variety of authors
including Fisher (1940), Hill (1974), and Williams (1952). An interesting
account of its historical development is given in Chapter 1 of Greenacre
(1984). The method is now primarily associated with the name Benzécri.

In its simplest form correspondence analysis of a two-way table is
equivalent to representing the observed table of counts, regarded as a
matrix N, by one with lower rank. If the independence model fits, we
represent the table by a matrix of fitted values of rank 1. Essentially the
method consists of finding the singular value decomposition of the
matrix, E, containing residuals from fitting the independence model, that

is

$$e_{ij} = \frac{n_{ij} - np_{i+}p_{+j}}{(np_{i+}p_{+j})^{1/2}} \tag{11}$$

where $p_{i+} = n_{i+}/n$, and $p_{+j} = n_{+j}/n$. The singular value decomposition of $\mathbf{E}$ consists of finding matrices $\mathbf{U}$, $\mathbf{V}$, and $\mathbf{\Delta}$ (diagonal), such that

$$\mathbf{E} = \mathbf{U}\mathbf{\Delta}\mathbf{V}' \tag{12}$$

where $\mathbf{U}$ contains the eigenvectors of $\mathbf{E}\mathbf{E}'$ and $\mathbf{V}$ the eigenvectors of $\mathbf{E}'\mathbf{E}$; $\mathbf{\Delta}$ contains the eigenvalues of $\mathbf{E}\mathbf{E}'$. Such a decomposition leads to

$$e_{ij} = \sum_{k=1}^{R} \delta_k^{1/2} u_{ik} v_{jk}, \qquad i = 1 \cdots r, \qquad j = 1 \cdots c \tag{13}$$

where r and c are the number of rows and columns of the table, R is the rank of the matrix $\mathbf{E}$, and u_{ik} and v_{jk} are the elements of the kth column of $\mathbf{U}$ and the kth column of $\mathbf{V}$, respectively; $\delta_1 \cdots \delta_R$ are the eigenvalues of $\mathbf{E}\mathbf{E}'$, so that

$$\text{trace}(\mathbf{E}\mathbf{E}') = \sum_{k=1}^{R} \delta_k = \sum_{i=1}^{r} \sum_{j=1}^{c} e_{ij}^2 = \chi^2 \tag{14}$$

where χ^2 is the usual chi-square statistic for testing independence.

Correspondence analysis essentially uses the first two columns of $\mathbf{U}$, $\mathbf{u}_1$, and $\mathbf{u}_2$, and the first two columns of $\mathbf{V}$, $\mathbf{v}_1$ and $\mathbf{v}_2$, to provide a graphical display of the residuals; the entries in $\mathbf{u}_1$ and $\mathbf{u}_2$ give the two-dimensional coordinates of points representing the rows of the contingency table; those in $\mathbf{v}_1$ and $\mathbf{v}_2$ give the coordinates for the column categories. Because

$$e_{ij} \simeq \delta_1^{1/2} u_{i1} v_{ji} + \delta_2^{1/2} u_{i2} v_{j2} \tag{15}$$

a large positive residual will occur when u_{ik} and v_{jk} for $k = 1, 2$ are large and of the same sign; a large negative residual will occur when u_{ik} and v_{jk} for each k are large and of opposite signs. When u_{ik} and v_{jk} are small or their signs are not consistent for each k, the residual will be small.

Applying the method to the data in Table 3 we obtain the eigenvectors and eigenvalues shown in Table 5; plotting the results gives Figure 1.

It is the graphical display resulting from a correspondence analysis that is its most attractive feature. For the present example a simultaneous

TABLE 5
Eigenvalues and Eigenvectors Arising from
Applying Correspondence Analysis to the Data in
Table 3[a]

Eye color	u_1	u_2	Hair color	v_1	v_2
LE	−0.535	−0.276	FH	−0.633	−0.521
BE	−0.327	−0.348	RH	−0.120	−0.064
ME	0.043	0.810	MH	−0.059	0.756
DE	0.778	−0.381	DH	0.670	−0.304
			BH	0.362	−0.245

δ_1	δ_2	δ_3	δ_4
1073.3	162.12	4.6	0.0

[a] See Table 3 for an explanation of the symobls LE, BE,
and so forth.

plot of the positions of eye colors and hair colors is obtained, which
allows a direct visualization of how eye colors are associated with
different hair colors. For example, it is clear from Figure 1 that a large
positive residual will occur when dark eyes are combined with black or
dark hair, and a large negative residual when combined with fair hair.

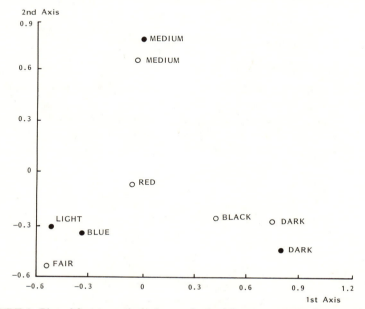

FIGURE 1. Plot of first two principal axes obtained from a correspondence analysis of the
data on eye color (●) and hair color (○) in Table 3.

Both latent class analysis and correspondence analysis can be expressed in a log-linear form. For example, Haberman (1979) shows that a latent class model for four categorical variables, i, j, k, and s is equivalent to the following log-linear model:

$$\log m_{xijks} = u + u_{1(x)} + u_{2(i)} + u_{3(j)} + u_{4(k)} + u_{5(s)} + u_{12(xi)} \tag{16}$$
$$+ u_{13(xj)} + u_{14(xk)} + u_{15(xs)}$$

where x is the underlying latent variable; the model in (16) implies that given x, the observed variables are independent.

Aitkin (1985) shows that correspondence analysis of a two-way table is equivalent to representing the association between rows and columns by a "multiplicative interaction" with a small number of terms, and that from a modeling point of view can be thought of as approximating the fit of the model

$$\log m_{ij} = u + u_{2(i)} + u_{2(j)} + \sum_{k=1}^{2} \delta_k^{1/2} u_{ik} v_{jk} \tag{17}$$

3. ILLUSTRATIVE COMPARISONS OF THE THREE APPROACHES

3.1. EXAMPLE 1: VOTING PATTERNS

3.1.1. Log-Linear Models

Because there are sampling zeros in these data, 0.5 was added to each cell before the analyses. The results of fitting three log-linear models—(1) main effects only, (2) main effects and first-order interactions, (3) main effects and first and second-order interactions—are given in Table 6. Payne (1977), who also fitted log-linear models to these data,

TABLE 6
Log-Linear Models for Voting
Behavior Data

Model	Effects	χ^2	df
1	One-variable	234.2	51
2	Two-variable	35.3	30
3	Three-variable	7.9	8

used forward selection techniques to search for a final model somewhere between models (2) and (3). The final model chosen was

$$\log m_{ijks} = u + u_{1(i)} + u_{2(j)} + u_{3(k)} + u_{4(s)} + u_{12(ij)} + u_{13(ik)} + u_{14(is)}$$
$$+ u_{24(js)} + u_{34(ks)} + u_{124(ijs)} + u_{134(iks)} \tag{18}$$

(where variable 1 is vote, variable 2 is sex, variable 3 is class, and variable 4 is age). This model implies the following:

- The association between any pair of variables vote, class, and age varies according to the level of the third ($u_{134} \neq 0$).
- Sex and class are not associated ($u_{23} = 0$).
- The association between sex and age does not vary with class ($u_{234} = 0$).
- The association between any pair of the variables vote, sex, and class does not vary with the level of the third ($u_{123} = 0$).
- There is no four-variable effect among the four variables ($u_{1234} = 0$).

3.1.2. Latent Class Analysis

The results of applying latent class analysis to the vote data are shown in Table 7. Examining the change in the goodness-of-fit statistic in going from 2 to 3 groups we do not find any substantial improvement in fit. The one-class model, which is equivalent to postulating the complete independence of the four variables, has, as we saw in the previous section, a chi-square value of 234 on 51 df. Consequently the two-class model is a substantial improvement over that of one class. The interpretation of the two-group solution seems relatively straightforward; class one consists predominantly of Labour voters from the working class, and class two of Tory voters coming in approximately equal proportions from the upper middle class, lower middle class, and working class. The age distribution and sex distribution in the two groups are similar, although there is a somewhat higher proportion of women than men among the Tory voters.

3.1.3. Correspondence Analysis

Because the vote data contain four variables, it is not altogether clear how correspondence analysis, as described in Section 2, may be applied. The first possibility to consider is *multiple correspondence analysis* (see Lebart, Morineau, & Warwick, 1984), which extends the

TABLE 7
Latent Class Analysis of Voting Behavior Data[a]

	Two-class solution		Three-class solution			Four-class solution			
	1	2	1	2	3	1	2	3	4
Proportion	0.57	0.43	0.31	0.44	0.25	0.34	0.25	0.25	0.16
P(C)	0.12	0.99	0.15	0.99	0.05	0.19	0.99	0.09	0.99
P(L)	0.88	0.01	0.85	0.01	0.95	0.81	0.01	0.99	0.01
P(M)	0.53	0.43	0.03	0.42	0.42	0.63	0.33	0.42	0.53
P(W)	0.47	0.57	0.37	0.58	0.58	0.37	0.67	0.58	0.47
$P(A_1)$	0.04	0.10	0.08	0.09	0.00	0.07	0.13	0.00	0.04
$P(A_2)$	0.24	0.32	0.24	0.31	0.24	0.25	0.36	0.24	0.25
$P(A_3)$	0.27	0.27	0.20	0.27	0.36	0.19	0.12	0.38	0.49
$P(A_4)$	0.29	0.21	0.30	0.22	0.38	0.30	0.28	0.27	0.11
$P(A_5)$	0.15	0.10	0.18	0.11	0.12	0.19	0.11	0.11	0.11
P(UMC)	0.10	0.31	0.06	0.31	0.13	0.08	0.29	0.11	0.36
P(LMC)	0.14	0.31	0.26	0.30	0.01	0.24	0.23	0.01	0.41
P(WC)	0.77	0.38	0.68	0.39	0.86	0.67	0.47	0.88	0.23
	$\chi^2 = 47.16$		$\chi^2 = 35.0$			$\chi^2 = 26.8$			
	42 df		33 df			24 df			

[a] See Table 2 for an explanation of the observations.

method to more than two classifying variables. Another possibility is to form a two-dimensional table by taking various combinations of the original four variables. It is this latter possibility that we consider here, and we first analyze the two-way table found by taking (sex × vote) as one variable and (class × age) as another. This is essentially equivalent to examining the residuals after fitting a log-linear model with the (sex × vote) and (class × age) marginals fixed. Such residuals contain information about interactions between the four variables *other* than the sex × vote and class × age interaction. The correspondence analysis of this table is shown in Figure 2. The first dimension in this diagram separates the working classes from the lower middle and upper middle classes, and the second dimension separates the Labour and Conservative voters; it clearly shows the association between class and voting behavior. The diagram also indicates that male Conservative voters are overrepresented in the upper middle class age Groups 2 and 3 and Conservative women occur proportionately more in the upper middle class age Group 1, and lower middle class age Groups 2 and 3. Interestingly they also occur more than would be expected among the older working class. Other tables could be constructed by crossing different pairs of variables

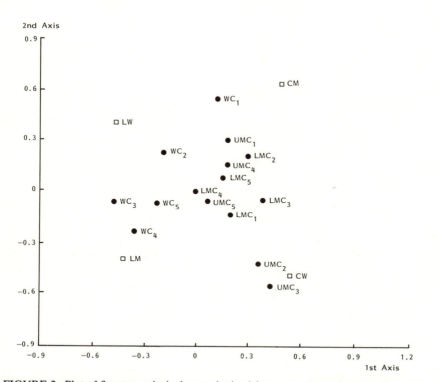

FIGURE 2. Plot of first two principal axes obtained from a correspondence analysis of the two-way table by cross-classifying (sex × vote) with (class × age). The data from which the cross classification is obtained are shown in Table 2. The following symbols are used: WC = working class; LMC lower middle class; UMC = upper middle class; CM = Conservative Party men; CW = Conservative Party women; LM = Labour Party men; LW = Labour Party women.

and this could prove a rich source for obtaining valuable insights into the data.

3.2. EXAMPLE 2: SUICIDE DATA

3.2.1. Log-Linear Models

The results of fitting a variety of log-linear models to these data are shown in Table 8. The data cannot be described adequately by any model simpler than the saturated. Here it might be more useful to analyze data from the two sexes separately. If this is done, tests of independence of

TABLE 8
Log-Linear Models for Suicide Data

Model	Variables	df	χ^2
1	(age) (sex) (method)	27	7954
2	(age sex) (method)	25	6640
3	(method sex) (age)	22	4299
4	(method age) (sex)	17	5169
5	(age sex) (method sex)	20	2985
6	(age sex) (method age)	15	3855
7	(age sex) (method age) (method)	10	128
8	(age sex method)	0	0

age and method of suicide give chi-square statistics for men and women of 1865 and 954, respectively, both on 10 df. These results indicate that there is a highly significant association between age and method in both sexes, but say little about the way the association differs between sexes as indicated by the significant second-order interaction term.

TABLE 9
Latent Class Analysis of Suicide Data[a]

	Two-class solution		Three-class solution		
	1	2	1	2	3
Proportion	0.42	0.58	0.38	0.37	0.25
P(M)	0.15	0.98	0.32	0.92	0.70
P(W)	0.85	0.02	0.68	0.08	0.30
$P(M_1)$	0.45	0.25	0.41	0.16	0.44
$P(M_2)$	0.01	0.07	0.00	0.07	0.10
$P(M_3)$	0.38	0.47	0.45	0.55	0.20
$P(M_4)$	0.02	0.11	0.01	0.10	0.13
$P(M_5)$	0.08	0.03	0.08	0.04	0.04
$P(M_6)$	0.06	0.07	0.05	0.07	0.09
$P(A_1)$	0.23	0.40	0.14	0.42	0.50
$P(A_2)$	0.55	0.46	0.62	0.43	0.41
$P(A_3)$	0.21	0.13	0.23	0.15	0.10
	$\chi^2 = 3022$		$\chi^2 = 2908$		
	18 df		9 df		

[a] See Table 1 for explanation of the abbreviations.

3.2.2. Latent Class Analysis

Applying latent class analysis to these data gives the results shown in Table 9. Such models clearly do not give a good fit to these data.

3.2.3. Correspondence Analysis

Initially a correspondence analysis was performed on men and women separately with the results shown in Figures 3 and 4. Figure 3 indicates that young men use Methods 1, 2, and 4 proportionately more than expected and Method 3 far less. The separate analysis of women in Figure 4 shows a roughly similar pattern. A more interesting and informative analysis is achieved by looking at a 6 × 6 table constructed by juxtaposing the two sexes. The correspondence analysis of this table is shown in Figure 5. The first axis separates men from women and shows among other things that young women use solids or liquids to kill themselves proportionately more than expected, whereas young men use gas rather more than expected. The diagram also clearly shows that all

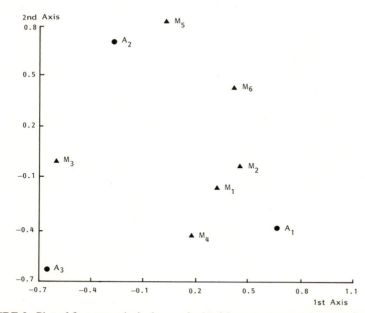

FIGURE 3. Plot of first two principal axes obtained from a correspondence analysis of the two-way classification (age group by cause of death) for males given in Table 1. The age groups are represented by A_1, A_2, and A_3. The methods of committing suicide by M_1, M_2, and so forth (see Table 1 for an explanation of their meaning).

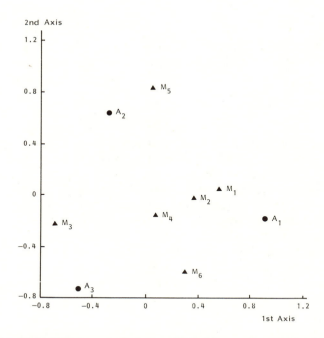

FIGURE 4. Plot of first two principal axes obtained from a correspondence analysis of the two-way classification (age group by cause of death) for females given in Table 1. The age groups are represented by A_1, A_2, and A_3. The methods of committing suicide by M_1, M_2, and so forth (see Table 1 for an explanation of their meaning.)

age groups of women use Methods 1 and 5 more than expected. The different relationship between method and age for men and women, as indicated by the significant second-order interaction in the log-linear analysis of these data, is clearly illustrated in this diagram.

It is tempting to interpret a method in association with an age group or sex/age group as in some sense "typical" of that age group, and a method distant from an age group as atypical; this is, however, potentially misleading. For example, in Figure 5, M_2 is close to males/Age Group 1; this only means, however, that young men use Method 2 *proportionately more* than other sex/age groups, not that they use predominantly Method 2; in fact they use Methods 1 and 3 far more often than Method 2.

4. DISCUSSION

Clearly the analysis of any set of data should be influenced by the substantive interests of the analyst. In many situations, however,

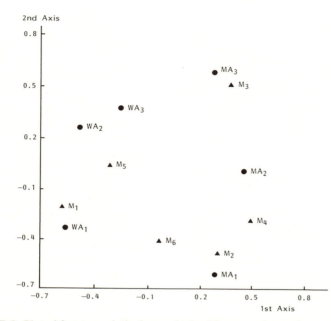

FIGURE 5. Plot of first two principal axes obtained from a correspondence analysis of the two-way table ([age × sex] by cause of death) obtained from consideration of both sexes in Table 1. The age groups for men are represented by MA_1, MA_2, and MA_3. The methods of committing suicide are represented by M_1, M_2, and so forth (see Table 1 for an explanation of their meaning).

particularly in the behavioral and social sciences, the questions that one might wish to ask of the data are often rather vague or ill defined. The way in which questions are asked will also depend on the philosophy of research worker or research team involved. French social scientists appear to be primarily interested in thinking of the patterns in their data in terms of abstract geometrical relationships that can be summarized by the graphical displays produced by correspondence analysis. Their English and American colleagues, however, seem to use methods much more appropriate to an empirical hypothesis-testing approach to research. Although it is possible to think of log-linear models in graphical or geometrical terms, this is rarely the view taken by their users. The use of log-linear modeling techniques almost always involves formal methods of statistical inference based on relative likelihoods. The users of correspondence analysis, on the other hand, are rarely interested in formal methods of inference. Indeed, Benzécri and his followers do not wish to base their decision on what they see as defective inferential methods based on questionable assumptions. They simply want to look at pictures and use their insight and/or experience to interpret them. Clearly we cannot rationally decide which of these two philosophies is "best".

Our approach is much more flexible, being more concerned with the use of data analytic methods to generate ideas. This is more like that advocated by Tukey (1977) and his followers. Here several methods of analysis might be applied to the data in the hope of discovering an interesting pattern or structure. In this paper we have considered three possible methods for the analysis of multidimensional categorical data. Our aim has not been to conclude that one method is better than another or that one of the methods is more appropriate for data sets of a particular type. Different exploratory techniques yield information about different aspects of the data. Of course, if the research worker is convinced of the necessity of formal methods of statistical inference, he or she will be more likely to use statistical modeling techniques rather than informal graphical displays. But there is no reason why one cannot use both. One can test ideas using log-linear models or latent class analysis and then illustrate the results through the graphical display generated by correspondence analysis or any other form of multidimensional scaling and/or cluster analysis. Alternatively one might wish to use correspondence analysis for an initial exploration of the data to generate ideas that can subsequently be tested through the use of an appropriate series of statistical models. One way of viewing our approach to the use of correspondence analysis in the illustrative examples above is as a method of exploration of patterns of residuals obtained by fitting log-linear models. Here the methods complement each other.

Latent class analysis is not so widely known or used as the other two methods, but it may in some cases provide an appealing way of describing the data. Our analysis of voting patterns (Table 2) illustrates this point. Other methods used for the classification of subjects and/or variables may also prove to be useful. It is interesting, for example, that French workers often used cluster analysis to supplement the results of correspondence analysis (Lebart, Morineau, & Warwick, 1984). Thinking of correspondence analysis as a special case of the general technique of multidimensional scaling (MDS) also leads one to the idea of the use of other forms of MDS. Would nonmetric MDS and the use of other distance measures between rows and columns of an appropriate contingency table yield useful and informative results?

ACKNOWLEDGMENT

This work was carried out as part of a collaborative Anglo-French project which is supported by the E.S.R.C. and the C.N.R.S.

5. REFERENCES

Aitkin, M. (1985). Paper presented at Workshop on Comparison of Correspondence Analysis and Statistical Modeling. University of Lancaster.

Benzécri, J. P. (1969). Statistical analysis as a tool to make patterns emerge from data. In S. Watanabe (Ed.) *Methodologies of pattern recognition* pp. 35–74. New York: Academic Press.

Birch, M. W. (1963). Maximum likelihood in three-way contingency tables. *Journal of the Royal Statistical Society, Series B, 25,* 220–233.

Bishop, Y. M. M., Fienberg, S. E., & Holland, P. W. (1975). *Discrete multivariate analysis: Theory and practice.* Cambridge, Massachusetts: MIT Press.

Butler, D., & Stokes, D. (1974). *Political change in Britain* (2nd ed.), London: Macmillan.

Dempster, A. P., Laird, N. M., & Rubin, D. B. (1977). Maximum likelihood from incomplete data via the EM algorithm. *Journal of the Royal Statistical Society, Series B, 39,* 1–38.

Everitt, B. S. (1977). *The analysis of contingency tables.* London: Chapman and Hall.

Everitt, B. S. (1984). *An introduction to latent variable models.* London: Chapman and Hall.

Everitt, B. S., & Hand, D. J. (1981). *Finite mixture distribution.* London: Chapman and Hall.

Fisher, R. A. (1940). The precision of discriminant functions. *Annals of Eugenics, 10,* 422–429.

Goodman, L. A. (1974). Exploratory latent structure analysis using both identifiable and unidentifiable models. *Biometrika, 61,* 215–231.

Green, B. F. (1951). A general solution of the latent class model of latent structure analysis and latent profile analysis. *Psychometrika, 16,* 151–166.

Greenacre, M. J. (1984). *Theory and applications of correspondence analysis.* London: Academic Press.

Haberman, S. J. (1979). *Analysis of qualitative data: Vol. 2. New developments.* New York: Academic Press.

Hill, M. O. (1974). Correspondence analysis: A neglected multivariate method. *Applied Statistics, 23,* 340–354.

Lazarsfeld, P. L., & Henry, N. W. (1968). *Latent structure analysis.* Boston, Massachusetts: Houghton Mifflin.

Lebart, L., Morineau, A., & Warwick, K. M. (1984). *Multivariate descriptive statistical analysis, correspondence analysis and related techniques for large matrices.* New York: Wiley.

Payne, C. (1977). The log-linear model for contingency tables. In C. A. O'Muircheartaigh & C. Payne (Eds.), *The analysis of survey data* (Vol. 2). London: Wiley.

Tukey, J. W. (1977). *Exploratory data analysis.* Reading, Massachusetts: Addison-Wesley.

Van der Heijden, P. G. M., & de Leeuw, J. (1985). Correspondence analysis and Complementary to Loglinear Analysis, *Psychometrika, 50,* 429–447.

Williams, E. J. (1952). Use of scores for the analysis of association in contingency tables. *Biometrika, 39,* 274–298.

CHAPTER SIX

A Latent Class Covariate Model with Applications to Criterion-Referenced Testing

C. MITCHELL DAYTON AND GEORGE B. MACREADY

1. INTRODUCTION

Modern developments of latent class models as pioneered by Lazarsfeld and Henry (1968) and extended by Proctor (1970), Goodman (1974, 1975), Haberman (1974, 1979), Dayton and Macready (1976, 1980a), and others have found a variety of applications in the social and behavioral sciences. An area of special interest has been criterion referenced testing, where latent class models offer an attractive and powerful alternative to latent trait models (Macready & Dayton, 1980). Latent class models can directly represent mastery/nonmastery status and have the advantage of permitting an objective determination of cutting scores for mastery classification (Macready & Dayton, 1977). Recently, there has been interest in extending latent class models to include information from concomitant variables, or covariates. Although categorical concomitant variables (i.e., variables used to group respondents) can be incorporated systematically into current latent class formulations (Clogg & Goodman, 1984, 1985, 1986; Macready & Dayton, 1980) and estimation carried out using available computer programs (e.g., Clogg, 1977), this chapter

C. MITCHELL DAYTON AND GEORGE B. MACREADY • Department of Measurement, Statistics, and Evaluation, College of Education, University of Maryland, College Park, Maryland 20742. The research was supported, in part, by NSF Grant No. SES-8318580 entitled "Development of Concomitant-Variable Latent Class Models."

develops a more general model in which latent class membership is functionally related to one or more categorical and/or continuous concomitant variables (see Dayton & Macready, 1980b for a restricted model of this type).

Although only relatively crude concomitant variable models can be fitted to real data at the present time owing to limitations in computing software and because of limited information concerning the identifiability of such models, these restrictions will certainly be overcome in the near future, and this class of models represents a theoretical extension that has promise for greatly extending the usefulness of latent class models. For example, information from previous performance (e.g., pretest scores) can be taken into account when fitting mastery models, and this should allow for more efficient determination of mastery/nonmastery status. Similarly, in applying hierarchical models to developmental data, age and/or sociometric variables could be incorporated in order to improve the prediction of stages of development. Furthermore, hypotheses concerning concomitant variables themselves can be tested. For example, one could assess whether or not sociometric status is related to latent developmental stages and whether, if related, it significantly improves classification compared to, say, only using age as a concomitant variable.

The remainder of this paper is divided into three sections. First, a general concomitant-variable latent class model is presented which can accommodate variables that are discrete, continuous, or some combination of discrete and continuous. Second, a specific restricted model that has applicability to criterion-referenced testing is introduced. This model assumes only two underlying latent classes (e.g., masters and nonmasters) and incorporates a logistic functional relation between the concomitant variables and latent class membership. The final section presents an application of a restricted form of the model to a set of educationally relevant data.

2. A GENERAL CONCOMITANT-VARIABLE LATENT CLASS MODEL

2.1. Notation and Model

Assume J random, dependent variables, Y_j, $j = 1, \ldots, J$ each of which is multichotomously scored, $r \in \{1, \ldots, R_j\}$, for a total of I cases ($i = 1, \ldots, I$). The observed response data per case may be arranged in a vector $\mathbf{y}_i = \{Y_{ij}\}$, $Y_{ij} \in \{1, \ldots, R_j\}$. In addition, assume K concomitant variables per case, X_k, $k = 1, \ldots, K$ with $\mathbf{x}_i = \{X_{ik}\}$ being the associated vector. Thus, the complete data record for a case involves the ordered pair, $(\mathbf{y}_i, \mathbf{x}_i)$. The concomitant variables are assumed to have a known distribution so that they are completely flexible and may include any combination of continuous and discrete measures. For example, a 0, 1 indicator variable might be used to represent sex of respondent, a continuous measure used to represent performance on a pretest, and a product variable used to measure the interaction of these two concomitant variables. Alternatively, a nonlinear relation involving a concomitant variable might be modeled by utilizing powers of the covariate.

Given $\mathbf{x}_i$, the probability of any observed response vector is assumed to be a sum of components arising from two or more (in general, L) latent classes:

$$P(\mathbf{y}_i \mid \mathbf{x}_i) = \sum_{l=1}^{L} \Theta_{l \mid \mathbf{x}_i} \prod_{j=1}^{J} \prod_{r=1}^{R_j} [\alpha_{jr \mid l\mathbf{x}_i}]^{\delta_{ijr}} \tag{1}$$

where δ_{ijr} is the Kronecker delta:

$$\delta_{ijr} = \begin{cases} 1 & \text{iff } Y_{ij} = r \\ 0 & \text{otherwise} \end{cases}$$

The latent class proportion, $\Theta_{l \mid \mathbf{x}_i}$, is conditional on $\mathbf{x}_i$, and the $\alpha_{jr \mid l\mathbf{x}_i}$ are recruitment probabilities, which are conditional on both the latent class and $\mathbf{x}_i$, associated with the jth dependent variable (i.e., variable j, response r, latent class l).

This model is the usual multichotomous response, latent class model written conditional on the concomitant variable, $\mathbf{x}_i$. For simplicity, it is assumed that the number of latent classes, L, is constant with respect to $\mathbf{x}_i$, although the model can be generalized so that different numbers of latent classes are postulated conditional on $\mathbf{x}_i$.

The model in equation (1) is based on the assumption of local independence, which justifies the multiplication of recruitment probabilities without concern for the specific order of responding to the dependent variables. Also, we require

$$\sum_{r=1}^{R_j} \alpha_{jr \mid l\mathbf{x}_i} = 1 \qquad \text{for all } j, l, \text{ and } \mathbf{x}_i \tag{2}$$

In addition, the probabilities associated with the latent classes are assumed to sum to 1 conditional on $\mathbf{x}_i$:

$$\sum_{l=1}^{L} \Theta_{l \mid \mathbf{x}_i} = 1 \qquad \text{for all } \mathbf{x}_i \tag{3}$$

In many applications in which the dependent variables are dichoto-
mously scored (i.e., $r \in \{1, 2\}$), the recruitment probabilities, $\alpha_{jr \mid l\mathbf{x}_i}$ can
be given interpretations in terms of "errors." For example, assume that
the response $r = 1$ is a success (i.e., a correct response) while the
response $r = 2$ is a failure (i.e., an incorrect response), and that latent
class 1 is comprised of masters and latent class 2 is comprised of
nonmasters. Then, conditional on $\mathbf{x}_i$, if $l = 1$, $\alpha_{j2 \mid 1\mathbf{x}_i}$ is the probability of
an omission error due to such factors as forgetting, fatigue, and so forth.
On the other hand, if $l = 2$, then $\alpha_{j1 \mid 2\mathbf{x}_i}$ is the probability of an intrusion
error due to factors such as guessing and specific knowledge.

As implied above, it is often relevant to interpret the latent classes as
representing definable types of hypothetical respondents. For example in
the context of mastery assessment, the idealized response vector made up
entirely of correct responses, $\mathbf{v}_1 = (1, 1, \ldots, 1)^T$, might be identified
with masters, while the idealized vector $\mathbf{v}_2 = (2, 2, \ldots, 2)^T$ made up
entirely of incorrect responses might be identified with nonmasters.

The unconditional probability of the dependent variable vector, $\mathbf{y}_i$,
at a point $\mathbf{y}_0$, say, can be derived by summing (when the concomitant
variables, X_k, are strictly discrete) or integrating (when the X_k are at
least partially continuous):

$$P(\mathbf{y}_0) = \sum_{\mathbf{x}} P(\mathbf{y}_0 \mid \mathbf{x}) \cdot P(\mathbf{x}) \qquad \text{for discrete } X \qquad (4a)$$

$$P(\mathbf{y}_0) = \int_{-\infty}^{+\infty} P(\mathbf{y}_0 \mid \mathbf{x}) \cdot f(\mathbf{x}) \, d\mathbf{x} \qquad \text{for continuous } X \qquad (4b)$$

The probability function, $P(\mathbf{x})$, in the discrete case or the density
function, $f(\mathbf{x})$, in the continuous case must be specified and may or may
not contain additional parameters that must be estimated from the data.
For example, in the continuous case, with $k = 2$ concomitant variables, it
could be assumed that the concomitant variables are jointly normally
distributed and, if the data are expressed in z-score form, only the
covariance parameter of the bivariate normal density would have to be
estimated. Similarly, for a discrete representation, the relative frequencies of occurrence of the cross-tabulation of the X_k in the sample
could be taken as defining the probability function and no additional
parameters would have to be estimated.

Although the case with at least some of the X_k continuous is of
theoretical interest, no generality is lost in practical applications if
attention is restricted to the strictly discrete case [equation (4a)]. By
conditioning the model on the observed values of the concomitant

variables, X_k, the definite Stieltjes integral in equation (4b) can be replaced by summation without loss of generality because realizations of the X_k can only involve the summational partition of the integral (Parzen, 1960). That is, whether or not, in theory, some or all of the concomitant variables are viewed as continuous, the actual sampled values will, of course, always be a discrete realization, and a convenient practice is to assume that generalization of results is restricted to exactly this realization of X_k. From a practical perspective, the discrete model in equation (4a) is more realistic to apply to actual data sets since no distributional assumption concerning the concomitant variables is needed if the sample proportions are utilized to determine $P(\mathbf{x})$. Thus, the remaining theoretical developments in this paper are conditioned on the discrete values of the concomitant variables.

2.2. SUBMODELS

The model in equation (1) is, in essence, an L-state latent class model written conditional on the values of $\mathbf{x}_i$, the observed vectors of concomitant variables. For J dichotomous dependent variables there are $JL + L - 1$ parameters to be estimated *for each unique value of* $\mathbf{x}_i$. Even assuming identifiability of the unrestricted model, it is apparent that parameter estimates could only be obtained for very large data sets. The model can be restricted in a variety of ways. One useful approach is to consider submodels for the latent class probabilities, $\Theta_{l|\mathbf{x}_i}$, and for the recruitment probabilities, $\alpha_{jr|l\mathbf{x}_i}$. Thus, the function

$$\Theta_{l|\mathbf{x}_i} = g(\mathbf{x}_i; \boldsymbol{\beta}) \tag{5}$$

can be set up where $g(\ )$ is a specified function and $\boldsymbol{\beta}$ is an associated vector of parameters, β_b, $b = 1, \ldots, B$. Since $\Theta_{l|\mathbf{x}_i}$ is defined on the 0, 1 interval, the function $g(\ )$ must map $\mathbf{x}_i$ onto this interval. If there are only two latent classes, then the probabilities of latent class membership are $g(\mathbf{x}_i; \boldsymbol{\beta})$ and $1 - g(\mathbf{x}_i; \boldsymbol{\beta})$. If there are three or more latent classes that are assumed to obey an hierarchical principle, then the *cumulative* probability of membership can be modeled by defining a series of functions: $g_1(\ )$, $g_1(\ ) + g_2(\ )$ and so forth, where scaling restrictions are imposed to preserve the 0, 1 bounds for $\Theta_{l|\mathbf{x}_i}$.

Similarly, the recruitment probabilities can be written in terms of submodels:

$$\alpha_{jr|l\mathbf{x}_i} = h_j(\mathbf{x}_i; \boldsymbol{\gamma}_j) \tag{6}$$

where it is assumed that a single functional form $h(\ )$ is specified but that each dependent variable, Y_j, has a unique vector of parameters $\gamma_j = \{\gamma_{jra}\}$, $a = 1, \ldots, A$. Of course, the function $h_j(\ )$ must map $\mathbf{x}_i$ onto the 0, 1 interval since this is the interval over which the recruitment probabilities are defined.

Through the use of submodels, the number of parameters to be estimated for the dichotomous response variable case is $B + JA$ assuming that the submodels are each written in terms of linearly independent parameters. Although an extensive treatment of specific submodels is beyond the scope of this chapter, some useful cases are as follows:

1. Set $\alpha_{jr \mid l\mathbf{x}_i} = \alpha_{jr \mid l}$ for all $\mathbf{x}_i$; thus, only the latent class proportions are functionally dependent upon the concomitant variables. A restricted model of this type is considered in Section 3, below.

2. Set $\Theta_{l \mid \mathbf{x}_i} = \Theta_l$ for all $\mathbf{x}_i$; thus, only the recruitment probabilities are functionally dependent upon the concomitant variables.

3. Assume that all of the X_k are dummy variables representing group membership. Then, the concomitant variable model is equivalent to simultaneously fitting an equivalent unrestricted latent class model to each separate group defined by the dummy variables. For example, suppose dichotomous response data are available for three separate age groups of respondents. If the concomitant variables, X_k, $k = 1, 2$, are two dummy variables representing membership in, say, the first two groups, then the general concomitant variable latent class model is equivalent to fitting an unrestricted latent class model to each of the three age groups. As cited in the "Introduction," there is some research literature dealing with the incorporation of grouping variables into latent class analysis (e.g., Clogg & Goodman, 1984, 1985, 1986; Macready & Dayton, 1980).

4. Set $\alpha_{jr \mid l\mathbf{x}_i} = \alpha_{jr \mid l}$ for all $\mathbf{x}_i$ and set $\Theta_{l \mid \mathbf{x}_i} = \Theta_l$ for all $\mathbf{x}_i$; thus, the concomitant variable model reduces to an ordinary unconstrained latent class model.

2.3. CLASSES OF HYPOTHESES

A great variety of hypotheses concerning the concomitant variable latent class model and associated submodels can be set up and might be appropriate in specific research situations. There are two approaches to hypothesis testing that arise in the context of latent class modeling. First, if standard errors for individual parameter estimates and covariances between pairs of estimates for different parameters can be estimated from the data (either exactly, asymptotically, or approximately using procedures such as the jackknife or bootstrap), then approximate, large-sample z statistics can be generated for hypotheses that equate parameters to preset values such as 0 or that equate pairs of parameters. And, second, approximate, large-sample chi-square tests can be utilized by assuming the hypothesis to be true in order to define a "reduced model." Then, both the unreduced (i.e., "full") and reduced models are fitted separately to the data, likelihood ratio chi-square goodness-of-fit statistics are calculated separately for each model, and the difference between the chi-square statistics is used to test the hypothesis (see Dayton & Macready, 1976).

Although there may be technical problems concerning the use of asymptotic likelihood ratio tests for testing hypotheses that equate parameters to "boundary" values (Aitkin, Anderson, & Hinde, 1981; Everitt & Hand, 1981), nevertheless, the difference chi-square procedure has been widely used and no satisfactory alternative is currently available. For "full" and "reduced" models as defined above, the parameters automatically obey the principle of subset inclusion in the sense that the parameters in the reduced model can be obtained from those in the full model by imposing one or more linear restrictions. However, subset inclusion will not be satisfied, in general, when it is desirable to compare alternate functional forms for submodels. For example, for a two latent class model and a single concomitant variable, the latent class proportion, Θ_{x_i}, might be modeled as a two-parameter logistic function,

$$\Theta_{xi} = [l + \exp(-\beta_0 - \beta_1 \cdot x_i)]^{-1}$$

or as a one-parameter cumulative exponential function,

$$\Theta_{x_i} = \beta \cdot \exp(-\beta \cdot x_i)$$

Although the logistic function is more complex in the sense of being based on a greater number of parameters, the exponential model cannot be derived from the logistic model by imposing linear restrictions. Although there have been some theoretical developments aimed at testing the relative fit of such nonconforming models (e.g., Atkinson, 1970; Cox, 1961, 1962; Efron, 1984), there is almost no literature concerning the practical implementation of these methods (but see Alvord & Macready 1985).

3. THE BINOMIAL-MIXTURE COVARIATE MODEL

3.1. THE CONSTRAINED MODEL

The general model in equation (1) along with the submodels in equation (4) allow the development of relatively simple to exceedingly complex concomitant variable models. The purpose of this section is to consider in detail a case that is relatively simple but that has applications to real-world analytic situations. Specifically, we assume just two latent classes, which may be identified with such dichotomous states as mastery/nonmastery, sick/well, and so forth. Thus, the latent class parameters can be written as

$$\Theta_{1 \mid x_i} \equiv \Theta_{x_i}$$

and

$$\Theta_{2 \mid x_i} \equiv 1 - \Theta_{x_i}$$

The probability of being in the mastery state is made conditional on the value of the concomitant variable, X. Further, we assume the discrete submodel of equation (4a) and use the two-parameter logistic function as the functional form in equation (5); thus we have

$$\Theta_{x_i} = g(x_i; \boldsymbol{\beta}) = [1 + \exp(-\beta_0 - \beta_1 \cdot x_i)]^{-1} \qquad (7)$$

where the parameters β_0 and β_1 are written without subscripts since there are only two latent classes. If the concomitant variable, X, is vector valued, then β_1 is an appropriately dimensioned vector of partial-slope coefficients. For simplicity, in the remainder of the development in this section, we assumed that all dependent variables are dichotomously scored and that there is only a single concomitant variable.

Finally, we assume that the recruitment probabilities are independent of the concomitant variable, x, so that equation (6) yields $\alpha_{jr \mid lx_i} = \alpha_{jr \mid l}$ for all x_i and we assume that there is a single "omission" error rate and a single "intrusion" error rate associated with all dependent variables (i.e., $\alpha_{j2 \mid 1} = 1 - \alpha_1$ and $\alpha_{j1 \mid 2} = \alpha_2$, say). The resulting model is equivalent to a mixture of two binomial processes with rate parameters α_1 and α_2. However, the mixture parameter, Θ_{x_i}, is dependent upon values of the concomitant variable, X. Assuming a positive degree of relation between membership in Latent Class 1 (e.g., "mastery") and the concomitant variable, we would expect Θ_{x_i} to be a nondecreasing function of X and the logistic function provides a potentially realistic representation for such relations.

Note that the binomial-mixture covariate model is based on a total of four parameters: the two binomial-rate parameters, α_1 and α_2 and the two logistic parameters, β_0 and β_1. The probability of a single observation can be written:

$$\begin{aligned}
P(\mathbf{y}_i \mid x_i) &= \Theta_{x_i} \cdot \alpha_1^{m_i}(1 - \alpha_1)^{J-m_i} \\
&\quad + (1 - \Theta_{x_i})\alpha_2^{m_i}(1 - \alpha_2)^{J-m_i} \\
&= \{[1 + \exp(-\beta_0 - \beta_1 \cdot x_i)]^{-1}\}\alpha_1^{m_i}(1 - \alpha_1)^{J-m_i} \\
&\quad + \{[1 + \exp(\beta_0 + \beta_1 \cdot x_i)]^{-1}\}\alpha_2^{m_i}(1 - \alpha_2)^{J-m_i}
\end{aligned} \tag{8}$$

where m_i is the number of "1" responses in the response vector $\mathbf{y}_i$. Because there is a constant "omission" error rate, $1 - \alpha_1$, and a constant "intrusion" error rate, α_2, the probability associated with an observation does not depend upon the ordered vector of responses, $\mathbf{y}_i$, but only on the count of "correct" responses.

3.2. NUMERICAL EXAMPLE

Before turning to the problem of estimating parameters and assessing fit of this model to actual data sets, consider in detail a fabricated example. Assume there are four dichotomous, dependent variables and that the binomial rate parameters are $\alpha_1 = .8$ and $\alpha_2 = .3$ for "masters" and "nonmasters," respectively. The data setting might be criterion-referenced testing where we are postulating that, on average, the latent class that has mastered the underlying concept has an 80% chance of responding correctly (and, of course, a 20% chance of committing an omission error). Similarly, the nonmastery class has, on average, only a 30% chance of responding correctly (which represents an intrusion error) and, thus, a 70% change of answering incorrectly, which is consistent with their lack of mastery. Ignoring any concomitant variable information, the masters and nonmasters will generate responses that each follow their own appropriate binomial distribution, and the overall distribution of responses will be a mixture determined by the relative numbers of the two groups in the population. Consider, now, a concomitant variable, X, which is related to latent class membership by means of the logistic function with $\beta_0 = -2$ and $\beta_1 = 1$:

$$\Theta_{x_i} = [1 + \exp(2 - x_i)]^{-1}$$

where the range of X is from 1 through 5 in discrete values. Table 1

TABLE 1
Latent Class
Proportions Conditional
on X

X	LC proportion
1	.27
2	.50
3	.73
4	.88
5	.95

shows the specific value for the latent class proportion associated with each value of X. Thus, at the lowest level of X, the proportion of "masters" is only .27, but this rises steadily to .95 at the highest level of X. Because X is known for each respondent, we have, in effect, five different binomial mixtures with mixing parameters ranging from values of .27 to .95 but with the same two binomial rate parameters, $\alpha_1 = .8$ and $\alpha_2 = .3$, for each of the mixtures. In order to assess the unconditional probabilities of the responses, it is necessary to know the probability function, $p(x_i)$, for the concomitant variable. In practice, the relative frequencies of X_1 through X_5 in the sample would be used as values of $p(x_i)$.

3.3. ESTIMATION PROCEDURES

The likelihood for a sample of size I can be written as the product

$$\Lambda = \prod_{i=1}^{I} P(\mathbf{y}_i \mid x_i) \tag{9}$$

where $P(\mathbf{y}_i \mid x_i)$, as given in equation (8), is defined on a casewise basis. Alternatively, it is common for concomitant variable data to be summarized in grouped data form. Assume there are q distinct values of the concomitant variable, $X_1, X_2, \ldots, X_q$. Then, for J dependent variables represented in terms of scores, $m = 0, 1, \ldots, J$, and joint frequencies of occurrence in the sample, f_{im}, the likelihood based on the grouped data distribution is

$$\Lambda = \prod_{i=1}^{1} \left\{ P(x_i) \sum_{m=0}^{J} C_m [\Theta_{x_i} \alpha_1^m (1 - \alpha_1)^{J-m} \right.$$
$$\left. + (1 - \Theta_{x_i}) \alpha_2^m (1 - \alpha_2)^{J-m}] \right\}^{f_{im}} \tag{10}$$

where C_m is the binomial coefficient for a score of m (i.e., the number of combinations of J things taken m at a time), and

$$P(x_i) = (1/I) \sum_{m=0}^{J} f_{im}$$

is the relative frequency of occurrrence for the ith level of the concomitant variable for $i = 1, \ldots, q$.

Maximum likelihood estimation is based on the likelihood in equation (10) with some ensuing simplifications made possible by utilizing the Fisher "method of scoring" (Rao, 1973). In the context of latent class analysis, these procedures are detailed in Dayton and Macready (1976). For the present case, a maximum-likelihood approach has been incorporated into FORTRAN computer programs (Dayton & Macready, 1987), which are available from the authors. In the next section, an empirical example is presented along with a consideration of various hypotheses that might be of interest.

4. APPLICATION OF THE BINOMIAL-MIXTURE COVARIATE MODEL

A sample of 258 American school children in Grades 6, 7, or 8 were given several short computational tests in a free-response format. The test chosen for analysis contained four exercises involving multiplication of mixed numbers with unequal divisors for the fractions involved. Before turning to the available concomitant variable information, it is useful to examine the data set in order to determine if it provides a reasonable candidate for latent class analysis. The frequencies associated with score groups, as well as expected frequencies and chi-square components for fitting an ordinary binomial distribution, are shown in Table 2. The binomial rate parameter is estimated to be .627 and it is apparent from the chi-square goodness-of-fit statistic of 141.72 with three degrees of freedom that a single binomial process is a very poor representation of the data. Furthermore, the major chi-square components are in the two extremes of the distribution, which suggests that a mixture may be a good choice for these data.

In addition to the computational test data, standard scores from the arithmetic computation section of the Metropolitan Achievement Test (MAT) were available on a per-case basis and subjects were classified by sex. The MAT scores were divided into seven levels using z scores rounded to whole numbers (i.e., -3 and less, -2, -1, 0, $+1$, $+2$, $+3$ or

TABLE 2
Fraction Data Fitted by a Single Binomial
Process

Score (k)	Observed frequency	Expected frequency	Chi-square component
0	25	4.99	80.24
1	38	33.58	.58
2	55	84.67	10.40
3	61	94.88	12.10
4	79	39.87	38.40
Total	258	257.99	141.72

more), and these levels were arbitrarily coded 0 through 6. The additional analyses involving concomitant variables were conducted in three stages:

1. A two-class model with constant intrusion and constant omission error rates across items was fitted to the total sample and separately to each sex group; in the notation of Formula 1, this model can be written:

$$P(\mathbf{y}_i) = \sum_{l=1}^{2} \Theta_l \alpha_l^{m_i} (1 - \alpha_l)^{J - m_i} \tag{11}$$

where, as before, m_i is the score value for the ith respondent.

2. For the total group and separately for the males and females, the appropriate parameter estimates for Θ_l, α_1, and α_2 from Stage 1 were used to generate expected frequencies at each level of the MAT concomitant variable.

3. The binomial-mixture, concomitant variable model of equation (7) was fitted to the total group and separately to the male and female samples.

Table 3 displays parameter estimates and results for chi-square tests for the noncovariate two-state latent class models. Each chi-square statistic is based on one degree of freedom because there are five score groups and three parameters are fitted to the data (i.e., df = 5 − 1 − 3 = 1). Although for the total sample the chi-square goodness-of-fit statistic is nonsignificant at conventional levels, the associated p value of .075 suggest that better-fitting models may be available. However, when the sex groups were considered separately, very good fit was attained for the males, while, for the females, fit was somewhat better than in the total sample. Also, we note that, overall 56% of the population is

TABLE 3
Results from Two-State Latent Class Models[a]

Statistic	Total sample	Males only	Females only
$\hat{\Theta}_1$ (masters)	.56 (.066)	.60 (.097)	.53 (.090)
$1 - \hat{\alpha}_1$	.34 (.044)	.29 (.068)	.38 (.056)
$\hat{\alpha}_2$	.15 (.030)	.18 (.045)	.12 (.040)
χ^2	3.165	.808	2.240
df	1	1	1
p value	.075	.359	.134

[a] Values within parentheses are estimated standard errors of the statistics that precede them.

estimated as being in a "mastery" state and that the intrusion error rate is estimated at .34 and the omission error rate at .15. Error rates of this magnitude are not unusual for tests of this type, although intrusion errors are somewhat high for a test in free-response format.

When these parameter estimates were used to generate expected frequencies at each level of the MAT concomitant variable, very poor fit was attained. In the total, male and female samples, respectively, the chi-square values were 146.73, 70.52, and 99.44. Each statistic is based on 25 degrees of freedom since there are $7 \times 5 = 35$ "cells" but three independent parameters are estimated, and the expected frequencies are constrained to sum to the level total for each of the seven levels (i.e., df $= 35 - 3 - 7 = 25$). Thus, although the model in Stage 1 is an adequate overall representation of the data, it provides very poor fit when the various subgroups based on MAT scores are considered.

Table 4 displays results from fitting concomitant variable models to the total group and to the sex subgroups. In each case, the model yields satisfactory fit, although fit is considerably better in the sex subgroups than in the overall sample. Note that the degrees of freedom are 24 because four parameters are estimated when fitting each concomitant variable model. The parameter estimates are remarkably consistent for the male and female groups, with intrusion error rates of .39 and omission error rates of .11 and .12 for males and females, respectively. Table 4 also shows the estimated proportion of masters at each covariate level. These values were generated by substituting the covariate values 0–6 in the appropriate logistic function (e.g., for the male sample, the function is $[1 + \exp(8.480 - 2.552x_i)]^{-1}$). This function reveals an exceptionally strong relation between MAT scores and mastery state with scores of 0–2 representing virtually all nonmasters and scores 4–6 representing virtually all masters.

TABLE 4
Results from Covariate Latent Class Models

Statistic	Total sample	Males only	Females only
Recruitment probabilities:			
$1 - \hat{\alpha}_1$	.39	.39	.39
$\hat{\alpha}_2$	.11	.11	.12
Logistic parameters:			
$\hat{\beta}_0$	−8.182	−8.480	−8.376
$\hat{\beta}_1$	2.603	2.552	2.825
LC property (masters) per covariate level:			
$\hat{\Theta}_0$	.00	.00	.00
$\hat{\Theta}_1$	.00	.00	.00
$\hat{\Theta}_2$	.05	.03	.06
$\hat{\Theta}_3$	.41	.30	.52
$\hat{\Theta}_4$	.90	.85	.95
$\hat{\Theta}_5$	.99	.99	1.00
$\hat{\Theta}_6$	1.00	1.00	1.00
Fit statistics:			
χ^2	34.406	24.840	27.121
df	24	24	24
p value	.078	.414	.299

In summary, this example illustrates the value of incorporating concomitant variables into latent class analyses. First, grouping subjects in terms of sex provided noticeably better fit when considering the two-class model. However, this model was unable to satisfactorily reproduce frequencies across the levels of the MAT. Finally, when MAT was included in the modeling by means of the concomitant variable model, fit to both the male and female samples was excellent and the parameter estimates indicated an intuitively satisfying model for the relation between MAT performance and latent class membership.

5. REFERENCES

Aitkin, M., Anderson, D., & Hinde, J. (1981). Statistical modelling of data on teaching styles. *Journal of the Royal Statistical Society (Series B)*, *144-4A*, 419–461.

Alvord, G., & Macready, G. B. (1985). Comparing fit of nonsubsuming probability models. *Applied Psychological Measurement*, *9*, 233–240.

Atkinson, A. C. (1970). A method for discriminating between models. *Journal of the Royal Statistical Society, 3,* 323–345.

Clogg, C. C. (1977). *Unrestricted and restricted maximum likelihood latent structure analysis: A manual for users* (Working Paper No. 1977-09). Pennsylvania State University.

Clogg, C. C., & Goodman, L. A. (1984). Latent structure analysis of a set of multidimensional contingency tables. *Journal of the American Statistical Association, 79,* 762–771.

Clogg, C. C., & Goodman, L. A. Simultaneous latent structure analysis in several groups. In N. B. Tuma (Ed.), *Sociological Methods 1985.* San Francisco: Jossey-Bass.

Clogg, C. C., & Goodman, L. A. (1986). On scaling models applied to data from several groups. *Psychometrika, 51,* 123–135.

Cox, D. R. (1961). Tests of separate families of hypotheses. In *Proceedings of the Fourth Berkeley Symposium on Mathematical Statistics and Probability, 1,* 105–123.

Dayton, C. M., & Macready, G. B. (1976). A probabilistic model for validation of behavioral hierarchies. *Psychometrika, 41,* 189–204.

Dayton, C. M., & Macready, G. B. (1980a). A scaling model with response errors and intrinsically unscalable respondents. *Psychometrika, 45,* 343–356.

Dayton, C. M., & Macready, G. B. (1980b). Concomitant-variable latent class models. Paper presented at Psychometric Society annual meeting, Iowa City.

Dayton, C. M., & Macready, G. B. (1987). MODMULT & MODBIN: Microcomputer Programs for Concomitant Variable Latent Class Models. *American Statistician, 41,* 238.

Efron, B. (1984). Comparing non-nested linear models. *Journal of the American Statistical Association, 79,* 791–803.

Everitt, B. S., & Hand, D. J. (1981). *Finite mixture distributions.* New York: Methuen.

Goodman, L. A. (1974). Exploratory latent structure analysis using both identifiable and unidentifiable models. *Biometrika, 61,* 215–231.

Goodman, L. A. (1975). A new model for scaling response patterns: An application of the quasi-independence concept. *Journal of the American Statistical Association, 70,* 755–768.

Haberman, S. J. (1974). Log-linear model for frequency tables derived by indirect observation: Maximum-likelihood equations. *Annals of Statistics, 2,* 911–924.

Haberman, S. J. (1979). *Analysis of qualitative data: Vol. 2. New developments.* New York: Academic Press.

Lazarsfeld, P. A., & Henry, N. W. (1968). *Latent structure analysis.* Boston: Houghton Mifflin.

Macready, G. B., & Dayton, C. M. (1977). The use of probabilistic models in the assessment of mastery. *Journal of Educational Statistics, 2,* 99–120.

Macready, G. B., & Dayton, C. M. (1980). The nature and use of state mastery models. *Applied Psychological Measurement, 4,* 493–516.

Parzen, E. (1960). *Modern probability theory and its applications.* New York: Wiley.

Proctor, C. H. (1970). A probabilistic formulation and statistical analysis of Guttman scaling. *Psychometrika, 35,* 73–78.

Rao, C. R. (1973). *Linear statistical inference and its applications.* New York: Wiley.

PART III

COMPARATIVE VIEWS OF LATENT TRAITS AND LATENT CLASSES

CHAPTER SEVEN

Test Theory with Qualitative and Quantitative Latent Variables

JÜRGEN ROST

1. INTRODUCTION

Latent trait models and latent class models can be considered as two alternative, mutually complementary approaches to analyzing data obtained from tests and questionnaires. Data of this type are characterized by the fact that a larger number of manifest variables, that is, the items, are observed and that all manifest variables refer to the same aspect of personality of the individuals and are designed to measure it. The ultimate aim of a test analysis is to make comparative statements about the individuals by representing them on a scale, that is, by allocating them values of a latent variable. In the case of *latent trait models,* this latent person variable is *quantitative,* the persons are represented on a metric scale. In the case of *latent class models* the latent person variable is *qualitative,* the persons are mapped into a set of categories or classes.

The *manifest* variables of test and questionnaire data have in many, if not in most, cases more than two categories and, moreover, usually *ordered* categories. Whenever rating response formats or a partial credit scoring rule for free answers is used, the resultant data are multicategorical and ordered. Whereas in latent trait theory a variety of models for ordered data have been developed recently (cf. Masters, Chapter 1, this volume), there are only few applications of latent class analysis (LCA) with ordered categories (e.g., Clogg, 1979). In particular, there is no

JÜRGEN ROST • Institute for Science Education (IPN), University of Kiel, D-2300 Kiel 1, Federal Republic of Germany.

suitable way of imposing restrictions on latent class (LC) models taking care and taking advantage of the order of categories. In an earlier paper (Rost, 1985) it was argued that equality constraints and parameter fixations are not suitable to impose order restrictions on LCA. The proposed binomial LC model for rating data, on the other hand, has been proved to be too restrictive in many respects, since it has only one parameter for each item and, owing to the independence assumption, the distributions of item responses are too steep (cf. Rost, 1985, in press). The main purpose of this chapter is to show an effective way of defining LC models for ordered data and to discuss three very general LC models for ordered categories that are generalizations of the dichotomous LC model and submodels of the unrestricted multicategorical LC model. Before this is dealt with in Sections 3 and 4, Section 2 discusses the distinction between traits and classes in the simple dichotomous case. These ideas will be generalized for more than two categories in Sections 4 and 5.

2. TRAITS VERSUS CLASSES

Usually the concept of *test theory* is associated only with latent trait models. In these models it is assumed that the probability of a particular item response, for example, a correct solution of an "achievement item" or a "yes" response on an attitude item, increases monotonously with the value of a latent variable of a person. Figure 1 shows various types of functional relationships between the response probability and the latent variable, which is termed *latent trait*. These curves, called *item characteristic curves* (ICC), differ in many respects. First, the function may be *continuous* and *strictly monotonous* as in the case of all logistic ICCs (cf. b, d, f, and g in Figure 1) and the normal ogive model (cf. h). Or it may be a discontinuous step function as in the case of the latent distance model, of which the Guttman and the Proctor models are special cases (cf. a, c, and e). Second, the ICCs may take account of *guessing probabilities*, that is, a lower asymptote below which the solution probability cannot sink (cf. c, d, e, and g). Third, the strength of the relationship between response probability and latent trait, represented by the *slope of the ICC*, may be assumed to vary from item to item (cf. f and g).

Other types of functional relationships play a minor role in latent trait theory, because they either have less favorable statistical properties, like the normal ogive function, which is almost indistinguishable from a logistic function (cf. h in Figure 1), or they presuppose too strong

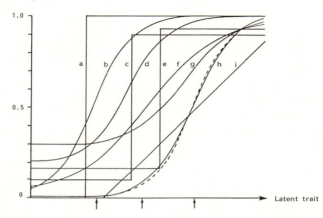

FIGURE 1. Different types of functional relationships between the response probability and the latent trait. The curves refer to the following models: a = Guttman scale (Guttman, 1950); b = Rasch model (Rasch, 1960); c = Proctor (1970); d = Keats (1974); e = Latent distance model (Lazarsfeld & Henry, 1968); f = Birnbaum model (Birnbaum, 1968); g = Three parameter logistic (Lord, 1980); h = Normal ogive (Lord & Novick, 1968) (the dashed line indicates the course of a logistic curve with appropriate parameters); i = Binomial model (Lord & Novick, 1968).

restrictions, such as, for example, constant item difficulties for all items in the binomial model (cf. i).

The proliferation of different latent trait models is drastically reduced by the demand that the test results, that is, the person measures, should be independent of the selection of items from a defined item domain. This demand, which Rasch (1960/1980, 1977) termed the demand for *specific objectivity*, implies that the ICCs of all items of an item domain have to be "parallel" in a particular sense. The corresponding model is obtained simply by equating the logit of the response probability, p_{vi}, to the sum of person ability, ξ_v, and item easiness, σ_i

$$\log(p_{vi}/(1 - p_{vi})) = \xi_v + \sigma_i$$

or

$$p_{vi} = \exp(\xi_v + \sigma_i)/[1 + \exp(\xi_v + \sigma_i)] \tag{1}$$

where v is the person index and i the item index (cf. Masters, Chapter 1, this volume, for a detailed introduction). This model and its generalizations for more than two categories will be the reference for latent trait analyses reported below. The following consideration, however, is not restricted to the Rasch model, but concerns all the latent trait models whose ICCs are shown in Figure 1.

All differences apart, these curves and step functions share in common that they describe the response probability as a monotonously increasing function of the quantitative latent variable. In order to lead over from latent traits to latent classes, it is now assumed that the quantitative latent variable has, whatever the reasons may be, a "discrete distribution" in a particular empirical context (cf. Clogg, Chapter 8, this volume). In other words, all persons of the sample can be grouped into a number of distinct classes, each class containing persons with the same value on the latent variable. In that case, the distribution is condensed on a particular number of "points" located on the latent continuum.

This is the idea of *located classes,* already used by Lazarsfeld and Henry (1968) to describe connections between latent classes and latent traits. Each located class can be characterized by the response probabilities of all items, because these are constant for all members of a particular class. The diagram showing the response probabilities of the items in a particular latent class is called "item profile." The important result here is that the item profiles of located classes do not intersect if the ICCs are monotonous, that is, if the ICCs follow any of the curves shown in Figure 1.

In Figure 2 the item profiles of three located classes are depicted. In this hypothetical example, the nine items of Figure 1 were taken and the arrows on the abscissa in Figure 1 indicate where the three classes are

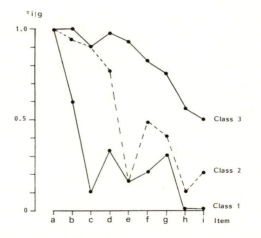

FIGURE 2. Item profiles of three located classes. The nine items were taken from Figure 1, where the location of the three classes on the latent trait is marked by arrows.

located. Hence, the response probabilities of Figure 2 were obtained by the intersection points of three vertical lines and the ICCs in Figure 1. In detail, Figure 2 illustrates that strictly monotone ICCs cause the item profiles neither to intersect nor to touch each other, whereas the weak monotone step functions of the latent distance items a, c, and e cause the item profiles to "stick together" whenever the classes are located at the same side of the step. In both cases, however, it can be said that they do not *intersect*.

A more exact definition of this relationship is that located classes are *ordered* whenever the ICCs are monotonous. According to Lazarsfeld and Henry (1968) latent classes are *ordered* if the classes can be numbered in such a way that the following inequality holds for all items i:

$$\pi_{i|1} \le \cdots \quad \pi_{i|g} \le \cdots \quad \pi_{i|h} \qquad (2)$$

The parameters $\pi_{i|g}$ are the conditional response probabilities of item i in class g according to the dichotomous latent class model (cf. Langeheine, Chapter 4, this volume)

$$p_{vi} = \sum_{g=1}^{h} \pi_g \pi_{i|g} \qquad (3)$$

where p_{vi} is the (unconditional) probability of person v to give a correct or "yes" response to item i, h is the number of classes, and π_g the class size parameter.

The relationship between monotonicity of the ICCs and the order of classes located on the latent continuum has twofold implications. If, on the one hand, a test or a questionnaire assumed to be unidimensional in the sense of a latent trait model produces unordered classes in a latent class analysis, the test obviously has nonmonotone ICCs, or the items may have monotonous ICCs but they are heterogeneous—they do not form a unidimensional test. Individual differences, in this case, may be represented better on a qualitative than on a quantitative person variable. If, on the other hand, the classes found in a latent class analysis are ordered, there is a good chance of fitting a latent trait model to these data and of representing individual differences on a quantitative variable.

This is illustrated by means of a real data example. Figure 3a shows the four-class solution of a set of knowledge items from a study on physics education (Häussler, Hoffman, & Rost, 1986). In contrast to the four classes of Figure 3b, those of Figure 3a are not ordered. A comparison of the two middle classes in Figure 3a shows that Class 2 has higher solution probabilities for items 6 to 10 whereas Class 3 has higher

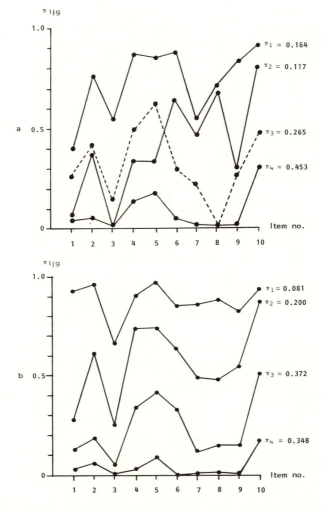

FIGURE 3. Item profiles of the four-class solutions (a) of the real data example and (b) of simulated, Rasch-homogeneous data.

probabilities for the first five items. In fact, the content of these items differs inasmuch as the correct answer to the first five items requires more "theoretical knowledge" while the other items require a knowledge gained by practical experience (cf. Häussler *et al.*, 1986). The data of Figure 3b, in contrast, were simulated on the basis of the Rasch parameters of the real data example and, hence, represent a homogeneous set of items. As expected, the classes are ordered.

How can it be tested statistically whether a latent trait model or a latent class model fits the data of both sets of items better? According to what has been said before, it should be expected that the real data set (cf. Figure 3a) is better fitted by a latent class model whereas the simulated data (cf. Figure 3b) are better fitted by a latent trait model.

Kelderman (1984) has shown that the dichotomous Rasch model can be formulated as a log-linear model and, hence, can be tested by means of the chi-square statistics used in the framework of log-linear and latent class modeling, that is, the likelihood ratio statistic

$$L^2 = 2 \sum_{\mathbf{x}} n(\mathbf{x})\{\log[n(\mathbf{x})] - \log[\hat{n}(\mathbf{x})]\} \qquad (4)$$

and the Pearson statistic

$$X^2 = \sum_{\mathbf{x}} [n(\mathbf{x}) - \hat{n}(\mathbf{x})]^2/\hat{n}(\mathbf{x}) \qquad (5)$$

where $n(\mathbf{x})$ is the observed, and $\hat{n}(\mathbf{x})$ the expected frequency of response pattern $\mathbf{x}$. Clogg (Chapter 8, this volume), Duncan (1984), and Kelderman (1984) give examples for comparisons of latent class models and the Rasch model reformulated as a log-linear model.

Andersen (1982; Chapter 9, this volume) also discusses model comparisons of latent trait and latent class models by means of chi-square statistics, but under the assumption that the latent variable of the Rasch model is normally distributed. However, to perform the goodness-of-fit tests (4) or (5) for a Rasch model it is neither necessary to estimate the complete parameter set of a log-linear model being equivalent to the Rasch model, nor to meet any distributional assumption about the latent variable. This point is stressed here because in the case of more than two response categories (cf. Section 4) parameter estimation via log-linear algorithms may become complicated (e.g., for 10 or 15 items with 5 or 7 categories) and distributional assumptions may not always be suitable.

Within the framework of Rasch models, the expected pattern frequencies $\hat{n}(\mathbf{x})$ can be obtained by conditioning the pattern probabilities $p(\mathbf{x})$ on the score variable $T = 0, 1, \ldots, t, \ldots, k$

$$\hat{n}(\mathbf{x})/N = p(\mathbf{x}) = p(\mathbf{x} \mid t)p(t) \qquad (6)$$

where N is the sample size and $p(t)$ can be estimated by means of the observed score frequencies, $p(t) = n(t)/N$ (cf. Bock & Aitkin, 1981). The conditional pattern frequencies are independent of the latent variable and are calculated by means of the symmetric functions $\gamma_t(\boldsymbol{\alpha})$ of the antilogarithmically transformed item parameters $\alpha_i = \exp(\sigma_i)$

$$p(\mathbf{x} \mid t) = \prod_i \alpha_i^{x_i} / \gamma_t(\mathbf{\alpha}) \qquad (7)$$

where $x_i \in \{0, 1\}$ is the ith component of the vector $\mathbf{x}$ (cf. e.g., Kelderman, 1984). Because the computation of the chi-square tests (4) and (5) using (6) requires the estimation of k independent score probabilities, in addition to the $k - 1$ independent item parameters, the statistics (4) and (5) are distributed with df $= 2^k - 2k$ degrees of freedom.

When these goodness-of-fit tests are to be applied to a larger number of items, for example, 10 in the example of Figure 3, the problem arises that many, if not most, expected pattern frequencies are close to zero and, hence, the asymptotic properties of chi-square statistics do not hold. Many authors, therefore, propose grouping patterns where the expected frequencies are too low, until a certain minimum value is reached or passed (cf. Andersen 1982; Chapter 9, this volume). While consensus should exist that the absolute minimum of expected frequencies is 1, some authors ask for a higher criterion value, e.g., 5 (cf. Bock & Aitkin, 1981).

The following results of the data example shown in Figure 3 are given for three different grouping criteria, that is, 1, 3, and 5 (cf. Table 1). It turns out that for the real data set with intersecting item profiles (cf. Figure 3a) the Rasch model and the three-class model do not hold, when any of the three grouping criteria is used. The four-class solution, on the

TABLE 1
Chi-Square Statistics (5) with Three Different
Grouping Criteria

Grouping criterion	Latent class model		Rasch model	
	Three classes	Four classes	Real data	Simulated data
χ^2	347	297	414	278
$\mathcal{E} > 1.0$ df	278	263	282	263
p	<0.01	>0.05	<0.01	>0.05
χ^2	158	129	206	126
$\mathcal{E} > 3.0$ df	113	98	123	114
p	<0.01	= 0.02	<0.01	>0.05
χ^2	104	76	155	89
$\mathcal{E} > 5.0$ df	63	50	76	70
p	<0.01	= 0.01	<0.01	>0.05

other hand, is non-significant at the 5% level, when all expectations are greater than 1, but comes close to the 1% boundary of significance when a more rigorous grouping is performed. Compared with the much higher values of the three-class model and the Rasch model, however, the fit of the four-class solution might be regarded as good enough.

In order to show that the decision between a qualitative and a quantitative latent variable is not affected by the drastic grouping procedure, the right column in Table 1 gives the results of the simulated, Rasch-homogeneous data, whose latent class structure was shown in Figure 3b. It turns out that the chi-square values are more or less of the same size as those of the four-class solution in the real data example.

This section was to show some relations of the Rasch model and LCA, treated from a test-theoretical point of view. The main purpose of this chapter is to put forward some new latent class models for ordered response categories. Response variables with ordered categories, obtained, for example, by rating response formats or partial credit scoring, probably represent the most frequent type of test and questionnaire data. It will turn out that the same parallelism between quantitative and qualitative latent variable models as discussed for the simple dichotomous case also holds for these polychotomous models.

3. THE THRESHOLD APPROACH IN LCA

Within the framework of Rasch models the threshold approach of formalizing multicategorical models has been proved a success and led to a series of models parametrizing the most important assumptions about ordered response formats (cf. Andrich 1978, 1982, 1985; Masters 1982). Masters (Chapter 1, this volume) is treating some of these models in detail. At first glance it may seem strange to transpose the threshold concept to LCA, because no latent continuum is given here, on which the thresholds could be located. The location of thresholds on a continuum is an essential part of the threshold concept, because otherwise it would make no sense to talk about "passing" or "not passing" a threshold.

Because earlier attempts at formalizing LC models for ordered categories were too restrictive and hence remained unsatisfactory (cf. Clogg, 1979; Rost, 1985), it is proposed to adopt the threshold concept for LCA. For this purpose quantitative latent variables are introduced, each quantifying the response tendency of a single latent class with respect to *a single item*. These latent variables can easily be obtained once the threshold probabilities are defined.

A threshold probability describes the probability of passing over to the next response category and thus presupposes that the preceding category has already been reached. Consequently, it is defined as a conditional probability, namely, the probability of responding in category x under the condition that the response is either in category $x - 1$ *or* in category x (cf., e.g., Masters, 1982):

$$\tau_{ixg} = p(r_{ig} = x \mid r_{ig} = x - 1 \text{ or } r_{ig} = x)$$

$$= \pi_{ix \mid g}/(\pi_{ix-1 \mid g} + \pi_{ix \mid g}) \tag{8}$$

where r_{ig} denotes a response of a person in class g on item i and $x = 0, 1, 2, \dots, m$. In order to facilitate further derivations, definition (8) has already been related to the model parameters of the polychotomous LCA

$$p(r_{vi} = x) = \sum_{g=1}^{h} \pi_g \pi_{ix \mid g} \tag{9}$$

where r_{vi} is the response of person v on item i [cf. (3)]. The threshold probability τ_{ixg}, therefore, denotes the probability of moving from response category $x - 1$ to category x for a person of latent class g responding to item i. It follows from definition (8) that the probability of category x, $\pi_{ix \mid g}$, can be calculated by means of the preceding response probability, $\pi_{ix-1 \mid g}$, and the threshold probability "between" these two categories:

$$\pi_{ix \mid g} = \pi_{ix-1 \mid g} \frac{\tau_{ixg}}{1 - \tau_{ixg}} \tag{10}$$

The response probabilities can also be written as a function of the threshold probabilities alone, but for this purpose, it is convenient first to introduce the latent variables mentioned above. This is done by linking the threshold probabilities to a latent dimension by means of the logistic function

$$\tau_{ixg} = \frac{\exp(\alpha_{ixg})}{1 + \exp(\alpha_{ixg})} \tag{11}$$

While τ_{ixg} is restricted to the interval $[0, 1]$, α_{ixg} is not. This is of some importance for incorporating restrictions on threshold distances into the model, because the values α_{ixg} can be treated as the "locations" of the thresholds on the latent dimension (cf. Figure 4a).

There are several reasons for choosing the logistic function in (11) (cf. Rost, 1987, in press), but one technical advantage becomes apparent immediately when inserting (11) into equation (10), namely, the reduction to the simple expression

$$\pi_{ix \mid g} = \pi_{ix-1 \mid g} \exp(\alpha_{ixg}) \qquad (12)$$

From (12) and the trivial condition

$$\sum_{x=0}^{m} \pi_{ix \mid g} = 1 \qquad (13)$$

it follows that the response probabilities depend on the threshold parameters in the following way (cf. Masters & Wright, 1984; Rost 1987):

$$\pi_{ix \mid g} = \frac{\exp(\sum_{s=0}^{x} \alpha_{isg})}{\sum_{x=0}^{m} \exp(\sum_{s=0}^{x} \alpha_{isg})} \qquad (14)$$

when $\alpha_{i0g} = 0$ is defined. Inserting (14) into the LC model (9) gives a reparameterization of the general LC model,

$$p(r_{vi} = x) = \sum_{g=1}^{h} \pi_g \frac{\exp(\sum_{s=0}^{x} \alpha_{isg})}{\sum_{x=0}^{m} \exp(\sum_{s=0}^{x} \alpha_{isg})}, \qquad \alpha_{i0g} = 0 \qquad (15)$$

which contains no kind of restrictive assumption at all. What is gained with the introduction of threshold parameters α_{ixg}, instead of probability parameters $\pi_{ix \mid g}$?

The answer is that order restrictions with respect to the response categories can hardly be defined in terms of the response probabilities $\pi_{ix \mid g}$. In particular, it is not true that ordered categories imply any order of their response probabilities. For example, a person with a mean attitude towards an attitude item with a rating response format has high probabilities in the middle categories and low at both ends of the rating scale. Rost (in press), therefore, argues that *unimodality* of the probability distribution of item responses would be an implication of ordered categories. Unimodality as a restriction for the latent class parameters $\pi_{ix \mid g}$, however, is hard to formalize without fixing a certain kind of distribution, for example, the binomial distribution (Rost, 1985).

As far as the threshold parameters α_{ixg} are concerned, it can be argued that ordered categories also imply an order of their thresholds (cf. Andrich, 1978). The rationale underlying this argument is that a transition from one category to the next must become harder, that is, less probable, with increasing category number, if the categories are really ordered. This does *not* mean that the response probabilities also decrease from category to category, but (cf. Figure 4) decreasing threshold parameters produce in any case a *unimodal* distribution of item responses (cf. Rost, in press).

When the threshold model (15) as a reparametrization of the LC model (9) is applied to a set of data, the order or nonorder of thresholds (and categories) is a *result* of parameter estimation and not an *a priori* assumption. But the threshold parameters α_{ixg} can be made subject to various restrictions, three of which are discussed in the following section.

4. THREE RESTRICTED LC MODELS FOR ORDERED CATEGORIES

The threshold parameters α_{ixg} introduced in the last section are specific for each item *i,* each class *g* and, of course, each category *x.* The three latent class models discussed in this section are obtained by fixing the threshold distances to be constant either across items, across classes, or across categories.

The *distances* of thresholds are made the object of restrictions, because the global response tendency of the individuals of a latent class towards a particular item should not be affected by the restrictions regarding the order of the categories. Hence another reparametrization of the threshold parameters α_{ixg} is done, namely, a split into their "mean location" on the latent trait and the distance of each particular threshold from this middle point (cf. Figure 4b). The location parameter λ_{ig} defined in this way,

$$\alpha_{ixg} = \lambda_{ig} + \beta_{ixg}, \qquad \sum_{x=1}^{m} \beta_{ixg} = 0, \qquad x > 0 \qquad (16)$$

quantifies what is called the *response tendency* of class *g* with respect to item *i,* because it is not only monotonously related to the threshold probabilities [cf. (11)]

$$\tau_{ixg} = \frac{\exp(\lambda_{ig} + \beta_{ixg})}{1 + \exp(\lambda_{ig} + \beta_{ixg})} \qquad (17)$$

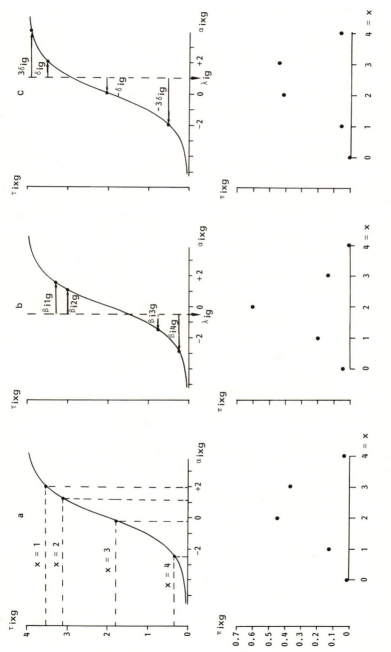

FIGURE 4. Threshold probabilities depending on the latent item variable and corresponding response probabilities (below). In 4b, threshold locations are parametrized as deviations form their mean; in 4c equidistant thresholds are depicted.

but also to the response probabilities [cf. (14)]:

$$\pi_{ix \mid g} = \exp\left(x \cdot \lambda_{ig} + \sum_{s=0}^{x} \beta_{isg}\right)\bigg/ d_{ig} \tag{18}$$

where $\beta_{i0g} = 0$ and the denominator d_{ig}, again, is simply the sum of the numerator over all categories, that is,

$$d_{ig} = \sum_{x=0}^{m} \exp\left(x\lambda_{ig} + \sum_{s=0}^{x} \beta_{isg}\right)$$

That means that, holding the distance parameters β_{ixg} constant, a higher location parameter λ_{ig} causes higher response probabilities in the upper categories and hence a higher expectation value of the response variable. Both are related monotonously.

The distance parameters β_{ixg}, on the other hand, can now be restricted in the way mentioned above. Equating them across all classes g, the following LC model is obtained:

$$p(r_{vi} = x) = \sum_{g=1}^{h} \pi_g \exp(x\lambda_{ig} + \varepsilon_{ix})/d_{ig} \tag{19}$$

where ε_{ix} is defined cumulatively, that is,

$$\varepsilon_{ix} = \sum_{s=0}^{x} \beta_{is}$$

with norming condition $\varepsilon_{i0} = \varepsilon_{im} = 0$ [cf. (16)] and d_{ig} is defined as above. In this model the threshold distances are not restricted across the items nor across the categories but they have to be the same for all persons, in all classes.

The second LC model is obtained by equating the distance parameters across the items,

$$p(r_{vi} = x) = \sum_{g=1}^{h} \pi_g \exp(x\lambda_{ig} + \psi_{xg})/d_{ig} \tag{20}$$

where ψ_{xg}, again, is defined cumulatively as

$$\psi_{xg} = \sum_{s=0}^{x} \beta_{sg}, \qquad \psi_{0g} = \psi_{mg} = 0$$

In this model no equidistance assumption is met, but the threshold distances have to be the same for all items. Furthermore, they are allowed to be different for each latent class.

The third model, finally, is an equidistance model because all thresholds of an item should have the same distances to their neighboring thresholds, but these distances may vary from item to item, and from class to class. It is convenient to parametrize only *half* the distance by a parameter δ_{ig}, so that the location of each single threshold is obtained by

$$\alpha_{ixg} = \lambda_{ig} - 2\left(x - \frac{m+1}{2}\right)\delta_{ig}$$

$$= \lambda_{ig} + (m - 2x + 1)\delta_{ig}$$

(21)

instead of (16). This is illustrated in Figure 4c. The resultant model is

$$p(r_{vi} = x) = \sum_{g=1}^{h} \pi_g \exp[x\lambda_{ig} + x(m - x)\delta_{ig}]/d_{ig}$$

(22)

because

$$\sum_{s=1}^{x} (m - 2s + 1)\delta_{ig} = [xm - (x^2 + x) + x]\delta_{ig}$$

In this parametrization, the distance parameter δ_{ig} is positive, when the threshold parameters and, hence, the threshold probabilities are decreasing, as it was decribed above to be a criterion for ordered categories.

In the following, the results of a real data set analyzed with the three LC models are described. In a study on environmental education in schools of the Federal Republic of Germany (cf. Eulefeld *et al.*, in press) a questionnaire, which was aimed at measuring something like the attitude toward environmental protection was answered by 422 teachers. The rating response format consisted of five categories from "don't agree" to "agree," including the neutral category " I can't decide." The items are given in Table 2.

First of all, the *structure* of the latent classes turned out to be quite stable under the different models. "Structure of latent classes" means here that the item profiles, that is, the expectations of the response variable in each latent class, are quite similar. At any rate, the differences are so small that the interpretation of the latent classes would not differ under the three models.

As Figure 5 shows, there is one group of teachers with a generally

TABLE 2

Items of the Questionnaire on Environmental Protection

1. Environmental protection is essentially the concern of the individual, for example, use of waste paper and glass recycling facilities, saving water, and energy.

2. Environmental protection primarily means caring for nature, for example, by setting up nesting boxes, creating ponds and biotopes.

3. The problems of environmental protection can only be tackled effectively at the national level, that is, especially by political means.

4. It is chiefly the responsibility of industry to curb pollution.

5. It is mainly the political parties that have great responsibility for environmental protection.

6. Efforts for environmental protection will be successful only when people learn to live more in touch with nature.

7. The problems of pollution are a consequence of the economic structure and can only be solved at this level.

8. The manufacture of nonpollutive products is mainly determined by consumer behavior.

9. Experts and scientists have primary responsibility for solving the problems of pollution.

high positive attitude towards means of environmental protection. A second group has relatively high ratings on the Items 1, 2, 6, and 8, whereas the third group rates the Items 3, 4, 5, and 7 higher than the second group. Comparing these two types of items, an interpretation is easily obtained. In the first item group mentioned, the responsibility for

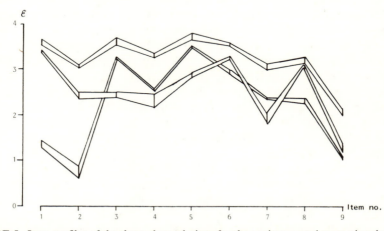

FIGURE 5. Item profiles of the three-class solutions for the environmental protection data. The two lines of each profile mark the minimum and maximum of the expectations of the response variable under the three LC models (19), (20), and (22).

TABLE 3
Log-Likelihoods and Number of Parameters in
Different LC and LT Models

Number of classes	Latent class model		
	(19) "ε_{ix}"	(20) "ψ_{xg}"	(22) "δ_{ig}"
2	−4985	−5020	−5059
	(46)	(25)	(37)
3	−4912	−4925	−4975
	(56)	(38)	(56)
4	−4882	−4870	−4908
	(66)	(51)	(75)

Latent trait model		
Partial credit	Rating scale	Dispersion
−4949	−5019	−5065
(71)	(47)	(53)

environmental protection is attributed to each individual person, whereas the other items ascribe an external responsibility for environmental protection, namely, to society, political parties, and the economic system.

This congruence of interpretation of the latent classes exists in spite of all differences regarding the parametrization of threshold distances. Table 3 shows the log-likelihoods of the three models with different numbers of classes, and the numbers of independent model parameters.

As can be seen, the equidistance model (22) fits the data worst, while the model with class-independent threshold distances (19) fits best. The parameters of this model show an interesting result regarding the threshold distances of the five-graded response format, which is obtained in the three-class as well as in the four-class solution. Table 4 gives the distances of the four-class solution and the obvious result is that the first and third distances are positive for all items, whereas the second is always negative.

Although it was argued in the previous section that decreasing threshold probabilities and therefore positive differences between their parameters,

$$\Delta(x, x + 1) = \alpha_{ixg} - \alpha_{i(x+1)g} \tag{23}$$

should be a criterion for ordered response categories, this particular

TABLE 4
Threshold Distances in the Four-Class
Solution of Model (19)

Item	$\Delta(1, 2)$	$\Delta(2, 3)$	$\Delta(3, 4)$
1	2.0	−2.0	2.0
2	2.7	−0.5	1.9
3	1.9	−1.2	1.2
4	1.4	−1.1	1.9
5	1.0	−0.9	1.5
6	0.6	−0.3	0.7
7	0.5	−0.1	1.0
8	0.7	−0.6	1.0
9	1.1	−0.2	1.2

result seems not to contradict the assumption of ordered categories, but says a lot about the neutral middle category of the response format ("I cannot decide"). Obviously a transition from the second to the third category and staying there is "harder" than moving on to the fourth category, causing the negative distance between the second and third threshold. This is only another description of the effect that most of the persons in this study tend to avoid the neutral middle category, regardless of where their response tendency towards a particular item is located. Figure 6 shows the influence of a negative distance between the two middle thresholds on the response probabilities for three different response tendencies.

In all cases the response probability of the third category is relatively

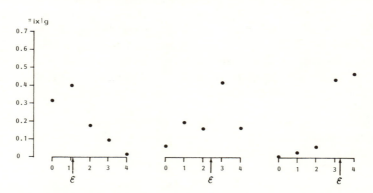

FIGURE 6. Three different response distributions for items with a negative distance between the Thresholds 2 and 3.

low, so that the distribution is unproportionally flat over this category. A bimodal distribution, however, is only obtained if there is a very high negative distance and the location parameter λ_{ig} is near zero, that is, the expectation of the response variable is about 2 (the number of the middle category).

That this effect is in fact independent of the class membership of the persons is confirmed by the parameters of model (20) with constant distances for all items. Here the middle distance is also negative in all latent classes of the three-class and the four-class solutions, while the first and third distances are always positive. From this result it becomes clear why the equidistance model (22) has the lowest likelihood of the three models under consideration (cf. Table 3).

Problems of testing the fit of these models and deciding for a number of classes are discussed in the next section, in connection with the comparison of latent trait (LT) and latent class models for ordered categories.

5. COMPARING THE FIT OF LT AND LC MODELS FOR ORDERED CATEGORIES

The LC models discussed in the previous section have, of course, their precursors in LT theory, and were developed with these models in mind. Nevertheless the relations are not quite symmetrical.

As a direct generalization of Rasch's dichotomous test model (1), Andrich (1978) developed the *rating scale model*

$$p(r_{vi} = x) = \exp(x\xi_v + x\sigma_i + \psi_x)/d_{vi} \tag{24}$$

d_{vi} denotes the sum of the numerator over all categories, and the *dispersion model* (Andrich, 1982)

$$p(r_{vi} = x) = \exp(x\xi_v + x\sigma_i + x(m - x)\delta_i)/d_{vi} \tag{25}$$

Masters (1982) derived the *partial credit model*

$$p(r_{vi} = x) = \exp(x\xi_v + \varepsilon_{ix})/d_{vi} \tag{26}$$

having both previous models as submodels (cf. Masters, Chapter 1, this volume). The correspondence of these models to the three LC models disussed in Section 4 is obvious. The asymmetry mentioned above lies in the fact that it is very easy to define the threshold distances as being dependent on the person variable within LC models. In Rasch models this is usually not the case, especially in any of the three models shown above.

Hence, a direct correspondence exists between model (19) and the partial credit model (26). The other two LC models (20) and (22), in contrast, have class-specific threshold parameters ψ_{xg} and δ_{ig}, and they are no submodels of (19). It is easy, however, to restrict them to submodels of (19) by equating their threshold parameters across all classes.

One way of evaluating the fit of a LC model and its corresponding LT model is the comparison of the *marginal* maximum likelihoods (MML) under both models (cf. Section 2). While these are automatically obtained with the application of LC models, because here the MML function is used for parameter estimation, the parameter estimates of Rasch models are commonly obtained by *unconditional* or *conditional* ML methods. The marginal likelihoods of Rasch models, on the other hand, require the computation of the symmetric functions $\gamma_t(\boldsymbol{\eta})$ of the item parameters $\eta_{ix} = \exp(\varepsilon_{ix})$, which may be cumbersome with many items and categories:

$$\log \text{MML} = \sum_{\mathbf{x}} n(\mathbf{x}) \left\{ \sum_{i=1}^{k} \varepsilon_{ix} - \log[\gamma_t(\boldsymbol{\eta})] \right\} + \sum_{t=0}^{m \cdot k} n_t \log[p(t)] \quad (27)$$

where

$$\gamma_t(\boldsymbol{\eta}) = \sum_{\mathbf{x}|t} \prod_{i=1}^{k} \eta_{ix}$$

and $\varepsilon_{ix} = \log(\eta_{ix})$ are the item parameters of the partial credit model (26). In the case of only two categories, recursive formulas for computing the symmetric functions are available (cf. Andersen, 1972). In the following, therefore, a generalization of these recursive functions for the item parameters of the partial credit model is given (cf. Rost, 1987). First the "symmetric function of order t without item i," $\gamma_t^{(i)}(\boldsymbol{\eta})$, is defined as

$$\gamma_t^{(i)}(\boldsymbol{\eta}) = \gamma_t(\boldsymbol{\eta}) - \sum_{x=1}^{m} \eta_{ix}\gamma_{t-x}^{(i)}(\boldsymbol{\eta}) \qquad \text{for } m \leq t \quad (28)$$

That is, from the symmetric function of order t all terms have to be subtracted, where any of the parameters of item i appears as a factor. For $t < m$ the sum in (28) runs only from $x = 1$ to $x = t$.

The symmetric functions of order t, then, can be computed by means of the functions (28) of a lower order (cf. Rost, 1987):

$$\gamma_t(\boldsymbol{\eta}) = \frac{1}{t} \sum_{i=1}^{k} \sum_{x=1}^{m} x\eta_{ix}\gamma_{t-x}^{(i)}(\boldsymbol{\eta}) \tag{29}$$

With the help of this formula, it is easy to compute the symmetric functions and the marginal likelihoods not only for the partial credit model. In order to apply formula (29) to the rating scale or the dispersion model, the partial credit parameters have to be replaced by their linear decompositions

$$\varepsilon_{ix}' = x\sigma_i + \psi_x \quad \text{or} \quad \varepsilon_{ix}' = x\sigma_i + x(m - x)\delta_i$$

The values of the MML functions of the data described in the previous section under the three LT models (24), (25), and (26) are shown in Table 3. The number of independent parameters that each likelihood is based upon equals $(k - 1) + (m - 1) + m \cdot k = 47$ for the rating scale model, $(k - 1) + k + m \cdot k = 53$ for the dispersion model, and $(m \cdot k - 1) + m \cdot k = 71$ for the partial credit model (cf. also Section 2). Compared with the likelihoods of the LC models (cf. Table 3), the fit of these LT models to the data is obviously poorer than the three- or four-class solutions of any of the LC models. This, of course, was to be expected, because each two of the item profiles of the latent classes intersect (cf. Figure 5) and, hence, the latent class structure is not compatible with the assumption that the classes are located on a latent continuum (cf. Section 2).

As far as statistical tests of the fit of a particular model are concerned [cf. (4) and (5) in Section 3], the problem of grouping patterns with too low expectations becomes even more serious for items with rating response formats. In the given example of nine items with five categories, for example, nearly 2 million different response patterns may be observed, but only 412 were observed (produced by 422 persons). Consequently a pattern grouping up to the minimum expected frequency of 1 (cf. Section 2) results in about 415 pattern groups containing, on the average, one person only.

If, despite all objections, the chi-square statistic (5) is computed according to this procedure, the conclusion drawn from the MML values is confirmed, namely, that the partial credit model has a better fit than the two-class models but a worse fit than the three- and four-class solutions (cf. Table 5). A statistical decision, however, does not seem possible, since all chi-square values are significant.

TABLE 5
Chi-Square Statistics (5) and Degrees of Freedom for
Pattern Groups with an Expectation Greater than 1

Number of classes	Latent class model			
	Unrestricted	(19) "ε_{ix}"	(20) "ψ_{xg}"	(22) "δ_{ig}"
2	$\chi^2 = 870$	$\chi^2 = 887$	$\chi^2 = 841$	$\chi^2 = 863$
	df = 346	df = 373	df = 395	df = 383
3	$\chi^2 = 557$	$\chi^2 = 562$	$\chi^2 = 586$	$\chi^2 = 741$
	df = 306	df = 361	df = 379	df = 361
4	$\chi^2 = 539$	$\chi^2 = 541$	$\chi^2 = 583$	$\chi^2 = 619$
	df = 270	df = 347	df = 366	df = 342

Latent trait model		
Partial credit	Rating scale	Dispersion
$\chi^2 = 692$	$\chi^2 = 875$	$\chi^2 = 910$
df = 343	df = 370	df = 363

It may be supposed that the far-reaching grouping procedure has an anticonservative effect on the Pearson statistic (5) and, hence, makes even the unrestricted four-class model significant. This supposition is supported by the observation that a higher grouping criterion (e.g., 3 or 5) relative to a lower one (e.g., 1) has the same effect (cf. Table 1). A systematic investigation of this phenomenon, however, has not yet been done.

On the other hand, a comparison of the chi-square values shown in Table 5 allows the conclusion that the fit of the LC models (19) and (20) is not significantly poorer than the fit of the unrestricted LC model. Moreover, the four-class solutions of these models do not seem to have a significant better fit than the three-class solutions. The chi-square differences between these models are evidently smaller than, for example, the differences to the two-class models, the LT models and the equidistance model (22). These conclusions, however, can only be drawn with the reservation that for this data example the global fit of the models could not be proven statistically, as it should be.

The problem of fitting a test model to data with a large number of patterns seems to be a general one and has not been solved satisfactorily yet (cf. Aitkin, Anderson, & Hinde, 1981). On the other hand, for Rasch models there is a variety of statistical tests (cf. van den Wollenberg, Chapter 2, this volume), which can—in a generalized form—be applied to multicategorical data in order to check the fit of a particular model. Some of the ideas underlying these tests can also be transferred to LC models, for example, the check whether the same latent class structure is obtained for different subgroups of items (cf. Rost, 1985). The primary goal of the MML statistics discussed here is an evaluation of models relative to alternative models and, especially, to provide a basis for deciding whether a quantitative or a qualitative latent variable represents individual differences better. As the examples have shown, this purpose is served by the MML statistics, despite the problem of statistical inference with larger item sets.

6. CONCLUSION

Whether test data are to be analyzed with a latent trait or a latent class model depends on the *psychological theory of individual differences* regarding those aspects of personality responsible for the test behavior. Many if not most psychological theories describe only qualitative individual differences, not quantitative ones. Why, then, should we not analyze tests and questionnaires in accordance with the underlying theory, that is, under the assumption of a categorical person variable? In other cases, a psychological theory of individual differences does not exist, but a relatively clear description of the *item domain* under consideration is available. If the items of a test constructed on the basis of such a theoretical reflection do not fulfil the strict requirements of homogeneity necessary for a metric scale level of the person variable, why do we not simply lower the scale level and use a categorical latent variable?

In the present chapter some new LC models were proposed, providing a basis for choosing between quantitative and qualitative person variables, not only in the case of dichotomous data, but also in the case of response variables with ordered categories. These LC models were constructed completely parallel to the most important LT models for ordered categories, namely, the rating scale, the dispersion and the partial credit model. In contrast to the latter, the LC models provide the possibility of defining threshold distances dependent on the person variable, have less restrictive requirements on item homogeneity, and show automatically, in the form of non-intersecting item profiles, when a LT model may be more adequate. A comparison of both types of models with respect to their fit to empirical data is possible by means of MML statistics.

As a conclusion, latent trait and latent class analysis can and should

be treated as two complementary ways of analyzing test and questionnaire data, be they dichtomous or multicategorical.

7. REFERENCES

Aitkin, M., Anderson, D., & Hinde, J. (1981). Statistical modeling of data on teaching
    styles. *Journal of the Royal Statistical Society, A, 144,* 419–461.
Anderson, E. B. (1972). The numerical solution of a set of conditional estimation
    equations. *Journal of the Royal Statistical Society, B, 34,* 42–54.
Andersen, E. B. (1982). Latent trait models and ability parameter estimation. *Applied
    Psychological Measurement, 6,* 445–461.
Andrich, D. (1978). Application of a psychometric rating model to ordered categories
    which are scored with successive integers. *Applied Psychological Measurement, 2,*
    581–594.
Andrich, D. (1982). An extension of the Rasch model for ratings providing both location
    and dispersion parameters. *Psychometrika, 47,* 105–113.
Andrich, D. (1985). An elaboration of Guttman scaling with Rasch models for measure-
    ment. In N. B. Tuma (Ed.), *Sociological Methodology 1985.* San Francisco: Jossey-
    Bass.
Birnbaum, A. (1968). Some latent trait models and their use in inferring an examinee's
    ability. In F. M. Lord & M. R. Novick (Eds.), *Statistical theories of mental test scores.*
    Reading, Massachusetts: Addison-Wesley.
Bock, R. D., & Aitkin, M. (1981). Marginal maximum likelihood estimation of item
    parameters: Application of an EM algorithm. *Psychometrika, 46,* 443–459.
Clogg, C. C. (1979). Some latent structure models for the analysis of Likert-type data.
    Social Science Research, 8, 287–301.
Duncan, O. D. (1984). Rasch measurement: Further examples and discussion. In C. F.
    Turner & E. Martin (Eds.), *Surveying subjective phenomena* (Vol. 2). New York:
    Russel Sage Foundation.
Eulefeld, G., Bolscho, D., Rost, J., & Seybold, H. (in press). *Praxis der Umwelterziehung
    in der Bundesrepublik Deutschland.* Kiel: IPN.
Guttman, L. (1950). The basis of scalogram analysis. In S. A. Stouffer, L. Guttman, E. A.
    Suchman, P. F. Lazarsfeld, S. A. Star, & J. A. Clausen (Eds.), *Studies in social
    psychology in world war II* (Vol. IV). Princeton, New Jersey: Princeton University
    Press.
Häussler, P., Hoffmann, L., & Rost, J. (1986). *Zum Stand physikalischer Bildung
    Erwachsener.* Kiel: IPN.
Keats, J. A. (1974). Applications of projective transformations to test theory.
    Psychometrika, 39, 359–360.
Kelderman, H. (1984). Log linear Rasch model tests. *Psychometrika, 49,* 223–245.
Lazarsfeld, P. F., & Henry, N. W. (1968). *Latent structure analysis.* Boston: Houghton
    Mifflin.
Lord, F. M. (1980). *Applications of item response theory to practical testing problems.*
    Hillsdale, N J: Erlbaum Associates.
Lord, F. M., & Novick, M. R. (1968). *Statistical theories of mental test scores.* Reading,
    MA: Addison-Wesley.
Masters, G. N. (1982). A Rasch model for partial credit scoring. *Psychometrika, 47,*
    149–174.

Masters, G. N., & Wright, B. D. (1984). The essential process in a family of measurement models. *Psychometrika, 49,* 529–544.

Proctor, C. H. (1970). A probabilistic formulation and statistical analysis of Guttman scaling. *Psychometrika, 35,* 73–78.

Rasch, G. (1977). On specific objectivity. An attempt at formalizing the request for generality and validity of scientific statements. *Danish Yearbook of Philosophy, 14,* 58–94.

Rasch, G. (1980). *Probabilistic models for some intelligence and attainment tests* (2nd ed.). Chicago: University of Chicago Press. (Original work published 1960).

Rost, J. (1985). A latent class model for rating data. *Psychometrika, 50,* 37–49.

Rost, J. (1987). *Probabilistische Testtheorie mit quantitativen und qualitativen latenten Variablen* (Avialable in polycop). Kiel: IPN.

Rost, J. (in press). Rating scale analysis with latent class models. *Psychometrika.*

CHAPTER EIGHT

Latent Class Models for Measuring

CLIFFORD C. CLOGG

1. INTRODUCTION

Most latent class analysis in contemporary social research is aimed at data reduction or "building clusters for qualitative data" (Formann, 1985, p. 87; see also Aitkin, Anderson, & Hinde, 1981). Some special restricted models in this area have of course been used to represent structural characteristics or behavioral processes (e.g., Clogg, 1981a; Goodman, 1974a). But a careful examination of the latent class models now available shows that none deal in a direct way with measurement, particularly if exacting standards are used to define how measurement should take place. Extensions and modifications of latent class models reported below are intended to remove this deficiency.

The term "measurement" here refers to the assignment of quantitative scores to subjects in a sample. The problem is that most concepts we would like to measure cannot be observed. Indirect observation—or indirect measurement—is the norm in the social sciences. Usually this refers to a process involving statistical analysis of models for multiple indicators, where the multiple indicators are assumed to contain the information necessary to create the measurements. The *measurement process* thus includes selection of indicators ("items"), model specification and validation, and estimation of parameters that serve as measurements.

To make matters concrete, consider the situation encountered in the

CLIFFORD C. CLOGG • Department of Statistics, Pennsylvania State University, University Park, Pennsylvania 16802. This research was supported in part by Grants Nos. SES-7823759, SES-8303838, and SES-8421179 from the National Science Foundation.

analysis of social surveys. The measurement process in this case is essentially as follows:

1. A set of items is assumed to result from indirect observation of a latent variable. This set consists of dichotomous variables (e.g., items with possible responses Agree and Disagree) and/or discrete-ordinal variables (e.g., items with possible responses Agree, Not Sure, Disagree).

2. A probabilistic model is formulated that (a) links the latent variable to the observed variables (as in a "system"), and (b) accounts for ("explains") the association among the observed items.

3. The model is tested and parameters are estimated. The parameter estimates are used to study the systemic properties of the variables (the general contours of which are already restricted by the model).

4. Each subject is assigned a value, or most probable value, on the scale represented by the latent variable.

Some comments on this set of procedures are useful. First, the number of items in the set of indicators is usually rather small in social surveys, say from two to six. This contrasts sharply with classroom tests for ability. Second, polytomous-nominal items can be used too, but this is seldom done. If we make allowance for "don't know" and "no answer" responses, however, polytomous items of this kind need to be considered (Clogg, 1984; Schuman & Presser, 1978). Third, probabilistic models are not always used. Cliff (1977) is a case in point, but many of the standard references on "scaling" seem to take deterministic models rather seriously. Whether classroom tests or social surveys are considered, however, I think that the "answering" or "responding" process is best captured by probabilistic models of some kind. In any event, statistical inference is greatly facilitated when stochastic models are used. Fourth, it is not enough to stop when a "good model" has been obtained and its parameter values are in hand. Many so-called measurement models stop short of measurement: most researchers seem to stop with the third step mentioned above. Fifth, true measurement models lead to a consistent strategy for within-group and across-group comparisons. These comparisons must be at least rankings, but within-sample comparisons should be based on interval-level measurements ("scores").

The agenda here is to discuss latent class models that can be used to produce measurements. Some familiarity with latent class analysis is assumed throughout our presentation. See Lazarsfeld and Henry (1968), Goodman (1978), Clogg (1981b), Clogg and Goodman (1984, 1985), and Formann (1985) for surveys and references. Three general types of latent models will be examined, each of which is related in one way or another to existing models. The *response-error model* is usually considered under the heading of scaling models (see Clogg & Goodman, 1986; Clogg & Sawyer, 1981). We show how specific models of this general kind can be used to create quantitative scores when a distributional assumption is supplied. The *latent-association model* is a modification of the usual latent class model that borrows ideas from association models for ordered data (cf. Goodman, 1984). The *scaled-latent-class* model (or the "parametric latent class model") resembles Rasch's (1960, 1966) model in several respects. Some of the interesting properties of Rasch's model apply to the scaled-latent-class model, but the problem of nuisance parameters does not arise. Before presenting these models and applying them to data, a brief digression on the axiom of local independence is included in order to motivate subsequent results.

2. LOCAL INDEPENDENCE

Let us assume for the moment that we have a set of J observable quantitative random variables, $Y_1, \ldots, Y_J$. The values that Y_j can take on are denoted by y_j. Let the set of possible values of y_j (the range space of Y_j) be denoted as ω_j, with ω_j consisting of points in the real line from a_j to b_j. Let $R_j = [a_j, b_j]$. Note that (a_j, b_j) need not equal $(-\infty, +\infty)$, nor do we require that ω_j consists of all points in R_j (continuity of Y_j is not necessary). Next suppose that a latent variable X with possible values x is measured by the set of the Y_j's. The range of x will be called ω_x; ω_x consists of points in the interval $R_x = [c_L, c_U]$. We do not require that ω_x consists of all points in R_x.

Now let $f(x) \geq 0$ denote the density function of X with $\int_{\omega_x} f(x)\, dx = 1$. Let $g_j(y_j) \geq 0$ denote the density function of Y_j with $\int_{\omega_j} g_j(y_j)\, dy_j = 1$. (This measure implicit here could be Lebesque measure, but this is not important.) Finally, let $h(y_1, \ldots, y_J, x)$ denote the joint density of $Y_1, \ldots, Y_J$ and X, with $h(\cdot) \geq 0$ for all arguments and

$$\int_{\omega_1} \cdots \int_{\omega_J} \int_{\omega_x} h(y_1, \ldots, y_J, x)\, dy_1 \cdots dy_j\, dx = 1$$

The *axiom of local independence* states that

$$h(y_1, \ldots, y_J, x) = f(x) \prod_{j=1}^{J} g_j(y_j \mid x) \tag{1}$$

for all $y_j \in \omega_j$, $j = 1, \ldots, J$, and all $x \in \omega_x$. This expression is the foundation of latent structure analysis, including factor analysis, latent trait analysis, and latent class analysis. In factor analysis, models are derived from this axiom by making certain assumptions. These are as follows:

1. The range spaces ω_j for the observed variables are assumed to be infinite $[\omega_j \in (-\infty, +\infty)]$. The range space for the unobserved variable X is also assumed to be infinite (*unrestricted range*).

2. All values in the real line are assumed possible for both y_j's and x (*continuity*).

3. The density $h(\cdot)$ is assumed to be a $(J + 1)$-dimensional normal. The densities $g_j(\cdot)$ and $f(\cdot)$ are thus univariate normal (*normality*).

4. Joint, marginal, and conditional distributions can be characterized by information contained in the first moments (means), second moments (or variances), and cross-moments (or covariances, or product-moment correlations) (*sufficiency of correlations*).

See Lawley and Maxwell (1967) and Joreskog and Sorbom (1979) for details. (Proponents of factor analysis rarely mention these assumptions, but factor analysis is based on them nevertheless.) Some of these assumptions are relaxed in the methods of Muthen (1979), Bartholomew (1980), and Bentler (1983).

In latent trait analysis, the Y_j's are discrete (dichotomous, discrete-ordinal, or polytomous-nominal), but X is continuous. The value of X for an individual is usually regarded as a parameter (see Andersen, 1980). In latent class analysis, the Y_j's and X are discrete (dichotomous, discrete-ordinal, or polytomous-nominal) (see Lazarsfeld and Henry, 1968). What all of these ostensibly different methods have in common is the reliance on the axiom of local independence (cf. Mooijaart, 1982). Some interesting implications of this axiom are given next.

If (1) holds, then any observed variable or any *subset* of observed variables can be *collapsed* without destroying the relationship. Let $h_{(1)}(y_2, \ldots, y_J, x)$ denote the marginal density obtained by integrating with respect to y_1, that is,

$$h_{(1)}(y_2, \ldots, y_J, x) = \int_{\omega_1} h(\cdot) \, dy_1 \tag{2}$$

Because $\int_{\omega_1} g_1(y_1 \mid x) \, dy_1 = 1$, from (1) we obtain

$$h_{(1)}(y_2, \ldots, y_J, x) = f(x) \prod_{j=2}^{J} g_j(y_j \mid x) \tag{3}$$

Local independence thus holds in the relevant marginal distribution if local independence holds in the original joint distribution. It is not difficult to see that local independence holds when any Y_j is collapsed or when any subset of the Y_j's is collapsed. An implication is that if local independence does not hold for any subset of the Y_j's, then it cannot hold for the complete set. A practical consequence of this fact is that random partitioning of the original set of items into subsets (perhaps disjoint subsets, but this is not required) can be used to test the axiom, assuming that a sufficient number of subsets is examined. A similar strategy for checking this assumption was advocated by Rasch (1966) in connection with his latent trait model. The point, however, is that this property stems from the form of the axiom, not from any particular *model* based on it. See Bishop *et al.* (1975, Chap. 2) and Fienberg (1980) for discussion of collapsibility conditions that are closely related to those above.

If (1) holds, then any observed variable can be *condensed* (grouped and/or truncated) without destroying the relationship of local independence. Without loss of generality, consider condensing Y_1 as follows. Divide the range space ω_1 into two disjoint sets ω_1' and ω_1'', where $\omega_1' \in [a_1, b_1']$ and $\omega_1'' \in [b_1', b_1]$. Next let y_1^* denote a value (any value) in ω_1''. Let $g_1'(y_1) = g_1(y_1)$ for $y_1 \in \omega_1'$ and let

$$g_1''(y_1^*) = \int_{\omega_1''} g_1(y_1)\, dy_1 \tag{4}$$

Now let $h'(y_1, \ldots, y_J, x) = h(y_1, \ldots, y_J, x)$ when $y_1 \in \omega_1'$; equation (1) holds whenever $y_1 \in \omega_1'$. Let $h''(y_1^*, y_2, \ldots, y_J, x)$ denote the density when $y_1 \in \omega_1'$ (i.e., when $y_1 = y_1^*$). We have the following expression for the joint density when $y_1 \in \omega_1''$:

$$h''(y_1^*, y_2, \ldots, y_J, x) = f(x) g_1''(y_1^* \mid x) \prod_{j=2}^{J} g_j(y_j \mid x) \tag{5}$$

In other words, local independence applies when Y_1 takes on values in ω_1' (by the original definition) and it applies when Y_1 takes on the value y_1^* in ω_1'' (or when y_1 values above a given point are condensed to the single value y_1^*). This result holds if the range space of Y_1 is divided into K disjoint subsets, with expressions like (5) holding for each modified density obtained from the grouping. Furthermore, the axiom holds if any subset of the Y_j's, or even all of the Y_j's, are condensed in any fashion (so long as the condensing does not result in degenerate random variables).

Some important observations on the condensability property are as follows. First, the converse of condensability is not true: if local independence holds for variables that have been grouped, it need not hold for the original variables. Grouping strategies are thus important for latent structure analysis, especially if the grouping cannot be checked against the "original" data. Second, *any* grouping will work if the axiom holds for the original variables. Of course, in sampling situations some groupings will be better than others, in the sense that some groupings will permit greater precision in estimation. Third, since any grouping of the Y_j's will still satisfy local independence, we can always think of latent trait or latent class models for discrete Y_j's as a representation of the axiom as it might hold for quantitative or continuous Y_j's that might have been available if our measuring instruments were more precise. Fourth, note that multinomial (or binomial) distributions will always describe the sampling distributions of grouped Y_j's. This fact leads to virtually all estimation procedures in latent class and latent trait analysis. The alternative seems to be an assumption of normality of some kind: this assumption is clearly violated for grouped Y_j's and it is probably violated for most sets of ungrouped Y_j's as well.

The above results pertain to collapsing or condensing Y_j's, the observable variables. They do not say anything about condensing latent variable X. In fact, if (1) holds, it is not possible in general to group or condense X. This is the case because x values appear in the conditional densities in (1). But it should be possible in many circumstances to divide the range space of X (ω_x) into disjoint subsets without altering the relationship of local independence very much. This would be the case if enough categories were used and if those categories were formed in the "right" way from the original X distribution. Of course, without knowing the distribution of X in advance we cannot use conventional grouping strategies (Cochran, 1983) developed for normal distributions.

Suppose that latent variable X is grouped into T categories (latent classes). Let x_t denote the value of x in the tth class of the grouped latent variable. That is, we partition the range space ω_x for X into disjoint sets $\omega_{x1}, \ldots, \omega_{xT}$, with ω_{x1} consisting of points in $R_1 = [c_L, c_1)$, ω_{x2} of points in $R_2 = [c_1, c_2), \ldots, \omega_{xT}$ of points in $R_T = [c_{T-1}, c_U]$. If $f(x)$ (the density function of X) were known, we might calculate the values x_t as follows:

$$x_1 = \int_{\omega_{x1}} xf(x)\,dx/F(c_1)$$

$$x_t = \int_{\omega_{xt}} xf(x)\,dx/[F(c_t) - F(c_{t-1})], \qquad t = 2, \ldots, T-1 \qquad (6)$$

$$x_T = \int_{\omega_{xT}} xf(x)\,dx/[1 - F(c_{T-1})]$$

Here x_t is the mean value of the distribution of X that is obtained by truncating below c_{t-1} and above c_t, with a similar statement applying to x_1 and x_T. $F(z)$ is the cumulative distribution function evaluated at z.

Now let $\pi_t^x = \int_{\omega_{xt}} f(x)\, dx$ denote the probability of the tth latent class. If T is sufficiently large and if the x_t values are suitably distributed over the range space ω_x, then

$$h(y_1, \ldots, y_J, x) \approx \pi_t^X \prod_{j=1}^{J} g_j(y_j \mid x_t), \qquad t = 1, \ldots, T \qquad (7)$$

As T becomes large, (7) is of course an identity, and this identity is the same as the axiom of local independence given in (1). When the Y_j's are grouped, or when they are discrete, the $g_j(y_j \mid x_t)$ values are conditional probabilities. [But note that the approximation in (7) holds even when the Y_j's are not grouped or are not discrete.]

Now consider the case where the Y_j's are discrete (perhaps through grouping). Suppose that Y_j takes on values k_j, with $k_j = 1, \ldots, K_j$. Let $\pi(k_1, \ldots, k_J, x_t)$ denote the joint probability $P(Y_1 = k_1, \ldots, Y_J = k_J, X = x_t)$, and let $\pi_{j(k_j) \mid t}$ denote the conditional probability $P(Y_j = k_j \mid X = x_t)$. The latent class model says that

$$\pi(k_1, \ldots, k_J, x_t) = \pi_t^X \prod_{j=1}^{J} \pi_{j(k_j) \mid t}, \qquad j = 1, \ldots, J \qquad (8)$$

Note that the right-hand sides of (7) and (8) are the same if the Y_j's are discrete. When both X and the Y_j's are discrete (with X having T classes and class values x_t, $t = 1, \ldots, T$), (8) and (1) are equivalent. When the X variable in (8) has been obtained by grouping, (8) is approximately the same as (1). The conventional definition of the latent class model does not use *scores* x_t, whereas the developments below all rest on the incorporation of these scores. Note that the conditional probabilities $\pi_{j(k_j) \mid t}$ could just as well have been written as $\pi_{j(k_j) \mid x_t}$.

The objective of latent class analysis in general is to find the number T of latent classes without taking account of scores x_t for the latent classes. By way of contrast, the approach presented here takes account of scores for the latent classes. By definition, these scores must be strictly monotonic (increasing or decreasing). And since these scores are viewed as points in the original range space ω_x for X—quantitative or continuous—we shall be interested in ways that this information can be taken into account. In short, we seek means by which the x_t values can be estimated and methods for assigning scores based on these values to individuals.

3. UNRESTRICTED LATENT CLASS MODELS

3.1. DATA

Table 1 contains three contingency data sets obtained from the 1982 General Social Survey. These contingency data sets are labelled A, B, and C, respectively. Data set A pertains to responses on four items on

TABLE 1

Three Cross Classifications of Respondents in the 1982
General Social Survey

Response pattern Item j =				Cross classification		
1	2	3	4	Free speech[a] (A)	Suicide[b] (B)	Satisfaction[c] (C)
1	1	1	1	571	105	287
2	1	1	1	11	0	27
1	2	1	1	64	1	46
2	2	1	1	19	0	19
1	1	2	1	67	4	127
2	1	2	1	18	0	21
1	2	2	1	18	4	46
2	2	2	1	30	3	33
1	1	1	2	66	10	242
2	1	1	2	14	1	38
1	2	1	2	36	3	77
2	2	1	2	39	3	37
1	1	2	2	63	62	191
2	1	2	2	53	16	49
1	2	2	2	49	444	117
2	2	2	2	270	724	134
Sample size (n)[d]				1388	1380	1491

[a] Items 1, 2, 3, and 4 refer to whether an antireligious person, a racist, a communist, or a militarist should be allowed to make a speech. Response 1 is "favor;" Response 2 is "not favor."

[b] Items 1, 2, 3, and 4 refer to whether suicide is acceptable when a person has an incurable disease, is tired of living (ready to die), has been dishonored, or has gone bankrupt, respectively. Response 1 is "yes;" Resonse 2 is "no."

[c] Items 1, 2, 3, and 4 refer to satisfaction with family, friends, hobbies, and residence, respectively. Response 1 refers to a "very great deal" or "a great deal;" Response 2 refers to "quite a bit," "a fair amount," "some," "a little," or "none."

[d] Number of respondents after deleting cases with "don't know" and "no answer" responses.

attitudes toward free speech—items which are very similar to those used three decades earlier by Stouffer (see McCutcheon, 1985). Data set B refers to responses on four items pertaining to whether suicide is acceptable in different circumstances. Data set C refers to responses on four items pertaining to satisfaction with various facets of life. See Clogg (1979) and Masters (1985) for analyses of similar items. For data sets A and B the original items were dichotomous, while for data set C the original items were dichotomized. See the notes to Table 1 for details. These data will be used for illustrative purposes later.

3.2. TWO-CLASS AND THREE-CLASS MODELS

Table 2 gives the likelihood-ratio chi-squared statistics (L^2) for the unrestricted latent class models that can be applied when $J = 4$ dichotomous items are considered. H_1 is the model of mutual independence among items, which can be viewed as a "one-class" model. The items are clearly not independent of one another, so latent class models with $T \geq 2$ are of interest.

Models H_2 and H_3 are the two-class and three-class latent structure models, respectively. The L^2 values indicate that neither model is satisfactory for data set A, both are satisfactory for data set C, and H_3 is satisfactory for data set B. Goodness of fit will not concern us here, however, since we are interested in the general status of these models as tools for measurement. The two-class model can in fact be used for measurement—in ways that are described next.

Table 3 gives the parameter values estimated for model H_2, that is, the latent class probabilities and the item conditional probabilities. Because $\pi_{j(1)|t} + \pi_{j(2)|t} = 1$, only the values $\hat{\pi}_{j(1)|t}$ are reported. The

TABLE 2
Chi-squared Values for Some Unrestricted Latent Class Models Applied to the Data in Table 1

Model	Description	Degrees of freedom	Likelihood ratio chi-square data set		
			A	B	C
H_1	Independence	11	1720.15	1328.65	363.22
H_2	Two-class model	6	67.29	56.60[a]	14.29
H_3	Three-class model	2	13.31	1.14	2.61

[a] One conditional probability $(\pi_{1(1)|1})$ has an estimated value of 1.00.

TABLE 3
Parameter Values for the Two-Class Model (H_2)

| | \multicolumn{6}{c}{Data set} |
| | \multicolumn{2}{c}{A} | \multicolumn{2}{c}{B} | \multicolumn{2}{c}{C} |
	$t = 1$	$t = 2$	$t = 1$	$t = 2$	$t = 1$	$t = 2$
$\hat{\pi}_t^X$	.58	.42	.09	.91	.66	.34
$\hat{\pi}_{1(1)\,|\,t}$	.98	.24	1.00^a	.41	.92	.45
$\hat{\pi}_{2(1)\,|\,t}$	.90	.24	.99	.06	.87	.25
$\hat{\pi}_{3(1)\,|\,t}$	.89	.18	.96	.01	.66	.24
$\hat{\pi}_{4(1)\,|\,t}$	.88	.14	.91	.01	.51	.20

[a] This estimate took on the boundary value of 1.00.

first latent class is associated with Response 1 on the items (i.e., "favor" for data set A, "yes" for data set B, "very satisfied" for data set C), while the second latent class is associated with Response 2 on the items. It can be observed that the latent class proportions are close to .5 for data set A, rather close to .5 for data set C, and very far from .5 for data set B. When we imagine that the dichotomous latent variable explicit in these models has been obtained by dichotomizing a variable with many more levels (perhaps even a continuous variable), the latent class probabilities tell us about the "threshold" point where the implicit dichotomization has taken place. This fact will be used later.

3.3. THE TWO-CLASS LATENT STRUCTURE AND THE COMMON FACTOR MODEL

The relationship between X and Y_j can be described in terms of the log-odds ratio,

$$\phi_{jx} = \log[(\pi_{j(1)\,|\,1}\pi_{j(2)\,|\,2})/(\pi_{j(1)\,|\,2}\pi_{j(2)\,|\,1})] \qquad (9)$$

(See Goodman, 1974a.) The estimates of these quantities for H_2 appear in column 1 of Table 4. The second column gives results from the common factor model. With the common factor model, the variables $Y_1 - Y_4$ are assumed to be dichotomizations of continuous variables $Y_1^* - Y_4^*$ which are normally distributed. The latent variable is also assumed to be continuous (call it X^*), and the joint distribution of $Y_1^* - Y_4^*$ and X^* is assumed to be normal. The tetrachoric correlations between pairs of items were factor analyzed using the procedure in LISREL version 6.1.

TABLE 4

The Relationship between Each Observed Item and the
Latent Variable under the Two-Class Latent Structure and
the Common Factor Model

	Log-odds ratio (H_2) $\hat{\phi}_{jx}$ (1)	Correlation (one-factor model) $\hat{\rho}_{j^*x^*}$ [a] (2)	Approximate correlation from $\hat{\phi}_{jx}$ of H_2 $\hat{\rho}_{jx}^*$ (3)
A			
Y_1	5.20	.94	.92
Y_2	3.31	.84	.82
Y_3	3.60	.86	.85
Y_4	3.80	.87	.86
B			
Y_1	∞	—[b]	1.00
Y_2	7.35	—	.92
Y_3	8.38	—	.92
Y_4	7.48	—	.91
C			
Y_1	2.61	.69	.71
Y_2	2.98	.76	.78
Y_3	1.81	.52	.58
Y_4	1.40	.41	.48

[a] The value of the likelihood-ratio statistic reported by LISREL was
136.13 for data set A and 14.34 for data set C, each with 2 df.
[b] LISREL could not calculate the solution for data set B.

Let $\rho_{j^*x^*}$ denote the correlation between X^* and Y_j^* under the common factor model. The estimated values of these correlations appear in the second column of Table 4 for data sets A and C; the solution for data set B could not be calculated. When columns 1 and 2 are compared, we see that an identical item ordering is obtained from the $\hat{\phi}_{jx}$ values for the two-class latent structure and the $\hat{\rho}_{j^*x^*}$ values for the common factor model (for data sets A and C).

Although the ranking of log-odds ratios $\hat{\phi}_{jx}$ and the ranking of the correlations $\hat{\rho}_{j^*x^*}$ are the same across items, the two sets of values are of course not directly comparable. Becker and Clogg (1986) give an approximate relationship between log-odds ratios and tetrachoric correlations that can be used in the present situation to draw comparisons. The precise form of this approximation is not of interest; but it adjusts the log-odds ratio for marginal skew in both dichotomous variables involved and employs a transformation like Fisher's z for the correlation coefficient. Let $\hat{\rho}_{j^*x}$ denote the correlation estimate obtained from $\hat{\phi}_{jx}$

using information in the two relevant marginal distributions (i.e., $\hat{\pi}_1^X$ and $\hat{\pi}_2^X$ for X and the observed marginal distribution for Y_j). These quantities appear in the third column of Table 4. Note the close agreement between the $\hat{\rho}_{j^*x^*}$ and the $\hat{\rho}_{jx}^*$: for data set A they agree to within .02; for data set C the maximum discrepancy is .07. In short, the parameters of the common factor model can be obtained from the parameters of the two-class latent structure for these tables. For data set B it is difficult to draw conclusions, since the common factor model based on tetrachoric correlations could not be estimated. (The common factor model using product-moment correlations can be estimated, but for data set B the product-moment correlations are very different from the tetrachoric correlations.)

3.4. Scores Based on the Two-Class Latent Structure

From the parameter estimates in Table 3, the *ordering* of the two latent classes in H_2 can be determined unambiguously. From either these values or the $\hat{\phi}_{jx}$ values in Table 4, the item ordering (say, in terms of "difficulty") can also be determined: the item difficulties can even be quantified in terms of either set of values. On the other hand, the two-class model does not incorporate information on the *distance* between latent classes. This information is required in order to complete the process of measurement. The simplest way to do this is to assign scores $x_t = t$, $t = 1, 2$. An alternative method is to conceive of X as a grouping of a latent variable X^* with known distribution. The latter strategy will be taken up further below when general response-error models are considered.

For simplicity, assume that $X^* \sim N(0, 1)$. Then X is obtained from X^* by dichotomizing at the "threshold point" c, where c is determined from the expression

$$\pi_1^X = F(c) \tag{10}$$

or $c = F^{-1}(\pi_1^X)$, where $F(\cdot)$ is the distribution function of the standard unit normal. The estimated c values are .210 for data set A, -1.357 for data set B, .423 for data set C. Using (6), with $T = 2$, gives x_1 and x_2 scores as

$$\begin{array}{lll}
\text{A:} & \hat{x}_1 = -.669, & \hat{x}_2 = .937 \\
\text{B:} & \hat{x}_1 = -1.818, & \hat{x}_2 = .174 \\
\text{C:} & \hat{x}_1 = -.550, & \hat{x}_2 = 1.087
\end{array}$$

These scores for the classes of X of course depend on the distributional assumption. Here a normal distribution was assumed, but any distributional assumption will work. (A simpler method, as noted earlier, is to let $x_t = t$.)

The measurement process is not completed until scores for individuals are calculated. A method for doing this is now discussed. Let a typical cell or response pattern in the contingency table be denoted as u, where u refers to a member of the set $\{(k_1, k_2, k_3, k_4), k_j = 1, 2\}$. Each cell u will in general be a mixture of latent classes 1 and 2. This mixture is determined by the probabilities $P(X = t \mid u)$. An average score for cell u can thus be determined by the relationship

$$s_u = \sum_{t=1}^{2} P(X = t \mid u)x_t \qquad (11)$$

TABLE 5
Response-Pattern Scores from the Two-Class Latent Structure
(Normality Assumption)

| Cell | | A | | | B | | | C | |
u	$\hat{X}^b$	$\hat{P}(\hat{X} \mid u)$	$\hat{s}_u$	$\hat{X}$	$\hat{P}(\hat{X} \mid u)$	$\hat{s}_u$	$\hat{X}$	$\hat{P}(\hat{X} \mid u)$	$\hat{s}_u$
1	1	1.00	−.67	1	1.00	−1.82	1	.99	−.53
2	1	.79	−.33	2	.97	.12	1	.88	−.35
3	1	.96	−.61	1	.98	−1.79	1	.83	−.27
4	2	.88	.75	2	1.00	.17	2	.73	.65
5	1	.95	−.59	1	.96	−1.74	1	.94	−.45
6	2	.91	.79	2	1.00	.17	1	.54	.21
7	2	.60	.29	2	.99	.14	2	.55	.35
8	2	1.00	.93	2	1.00	.17	2	.94	1.00
9	1	.94	−.57	1	.98	−1.78	1	.96	−.48
10	2	.92	.81	2	1.00	.17	1	.64	.04
11	2	.65	.37	2	.96	.10	1	.55	.19
12	2	1.00	.93	2	1.00	.17	2	.92	.95
13	2	.71	.47	2	.99	.15	1	.79	−.22
14	2	1.00	.93	2	1.00	.17	2	.78	.72
15	2	.99	.91	2	1.00	.17	2	.83	.81
16	2	1.00	.94	2	1.00	.17	2	.99	1.06

[a] The "percent correctly allocated" into latent classes was 95.9%, 99.9%, and 87.2% for data sets A, B, and C, respectively. The corresponding lambda measures of association between X and the items were .90, .99, and .62.
[b] $\hat{X}$ refers to predicted latent class; $\hat{P}(\hat{X} \mid u)$ refers to the probability of $\hat{X}$ given Cell u; $\hat{s}_u$ is the score for Cell u.

where the x_t scores for latent classes are determined here from a distributional assumption. Table 5 gives the predicted latent class, the estimated probability of this class, and the scores $\hat{s}_u$ for all cells in the three data sets. For example, Cell 7 [response pattern $(1, 2, 2, 1)$] in data set A has $\hat{P}(X = 2 \mid u = 7) = .60$, so $\hat{s}_7 = .40(-.669) + .60(.937) = .29$. (All calculations were carried out to more decimal places than indicated in the table.) To summarize, a distributional assumption applied to the two-class latent structure *a posteriori* was used to "scale" the latent classes and produce measurements for all individuals.

It should be noted that the above results were obtained for the two-class latent structure. The two latent classes in model H_2 can be ordered (e.g., as "High" and "Low"), and this fact enabled us to apply a distributional assumption. It is incorrect, however, to apply this same logic to unrestricted three-class models for $J = 4$ items for two reasons: (a) three-class models with unrestricted parameter values are not identifiable (Goodman, 1974b; Lazarsfeld & Henry, 1968), and (b) the three latent classes in a three-class model *cannot be ordered* unambiguously on either *a priori* or *a posteriori* grounds. Unrestricted latent class models with $T \geq 3$ are consistent only with a nominal-level X classification, which makes measurement very difficult. The models presented next overcome this limitation.

4. THE GENERAL RESPONSE-ERROR MODEL

4.1. Scaling Models Have Ordered Latent Classes

Let us assume now that the *item ordering* (say, in terms of difficulty) is given. Item marginals are often used to determine the ordering (see Clogg & Sawyer, 1981). The observed item marginals for Y_1-Y_4 in the 1982 General Social Survey were used to order the items in Table 1. (This ordering was consistent with the ordering inferred from marginals in the 1977 General Social Survey.) Under Guttman's (1950) deterministic model of scaling, the following scale-type response patterns are obtained:

$$S = \{(1, 1, 1, 1), (1, 1, 1, 2), (1, 1, 2, 2), (1, 2, 2, 2), (2, 2, 2, 2)\} \quad (12)$$

Note that these correspond to cells $u = 1$, 9, 13, 15, and 16 of the original tables. We now number these scale-type response patterns as $t = 1, \ldots, 5$; note that scale type t has $5 - t$ "1" responses. Note further that these scale types are *ordered*: the tth scale-type response pattern indicates "more" of the trait than the $(t + 1)$th scale-type response pattern.

Response-error models of the kind considered here posit $T = 5$ latent classes. (For J items there will be $J + 1$ latent classes, corresponding to the number of scale-type response patterns.) Members of the tth latent class have an expected response pattern equal to that of scale type t. The latent classes of X are thus *ordered,* because the scale types in (12) are clearly ordered. Now let α_{jt} denote the probability of a response error on the jth item for an individual in the tth latent class (i.e., given that an individual has expected response pattern equal to that of the tth scale type). For example, consider an individual in Latent Class 2, with expected response pattern $(1, 1, 1, 2)$. The response pattern $(2, 1, 2, 2)$ would be observed for such an individual with probability $\alpha_{12}(1 - \alpha_{22})\alpha_{32}(1 - \alpha_{42})$, representing error responses on Items 1 and 3 and correct responses on Items 2 and 4. It will be noted that α_{jt} is either $\pi_{j(1)|t}$ or $\pi_{j(2)|t}$ in a five-class latent structure. The five-class latent structure is not identifiable without restricting the parameters in some fashion. In other words, it is not possible to posit an ordered X classification without imposing some restrictions. Thus, the α_{jt} values (or the $\pi_{j(k_j)|t}$ values) must be restricted in order to obtain meaningful models.

4.2. Results Obtained from Some Response-Error Models

Table 6 presents L^2 values for the response-error models that will be considered here. H_4 is Proctor's (1970) model obtained by restricting $\alpha_{jt} = \alpha$ for $j = 1, \ldots, 4$ and $t = 1, \ldots, 5$. H_5 is the *item-specific* model, $\alpha_{jt} = \alpha_{j.}$; H_6 is the *type-specific* model, $\alpha_{jt} = \alpha_{.t}$; H_7 is Lazarsfeld's latent distance model, which allows error rates to depend on both the items and

TABLE 6
Chi-Squared Values for Some Response-Error Models

Model	Description	Degrees of freedom	Likelihood-ratio chi-square data set		
			A	B	C
H_4	Proctor's model, $\alpha_{jt} = \alpha$	10	105.81	20.16	54.27
H_5	Item specific, $\alpha_{jt} = \alpha_{j.}$	7	53.55	10.24	8.55
H_6	Type specific, $\alpha_{jt} = \alpha_{.t}$	6	47.60	7.58	29.51
H_7	Latent distance	5	38.71	4.57	5.56

the latent classes in certain specified ways. (See Clogg & Goodman, 1986; Clogg & Sawyer, 1981; and Rindskipf, 1983 for details.) We see from the L^2 values reported in Table 6 that none of these models fit well for data set A, that models H_5–H_7 are satisfactory for data set B, and that models H_5 and H_7 are satisfactory for data set C.

4.3. Scores for X and for Response Patterns

The estimated latent class probabilities $(\hat{\pi}_t^X)$ for models H_4–H_7 appear in Table 7. These can be regarded as relative frequencies of an ordinal-level variable X, since X is ordered by the one-to-one correspondence between latent classes and scale types. Now assume that X has been obtained by grouping a quantitative (or continuous) latent variable X^*, much as in the previous section. We seek scores x_t, $t = 1, \ldots, 5$, and as before a distributional assumption seems to be required to obtain them. If X^* has density $f(x)$ and distribution function $F(\cdot)$, then (6) can be applied. The threshold values c_t, $t = 1, \ldots, 4$, are determined from the expression $F^{-1}(c_t) = \pi_1^X + \cdots + \pi_t^X$. The estimated threshold values and the estimated scores $\hat{x}_t$ appear in Table 8 for the latent distance model (H_7 in Tables 6 and 7), assuming that $X^* \sim N(0, 1)$. The formula

$$s_u = \sum_{t=1}^{5} P(X = t \mid u)x_t \qquad (13)$$

TABLE 7

Estimated Latent Class Probabilities under the Models in Table 6

Model	Data set	Estimated latent class probabilities				
		$\hat{\pi}_1^X$	$\hat{\pi}_2^X$	$\hat{\pi}_3^X$	$\hat{\pi}_4^X$	$\hat{\pi}_5^X$
H_4 (Proctor)	A	.56	.01	.06	.02	.32
	B	.08	.01	.04	.33	.54
	C	.33	.21	.20	.10	.16
H_5 (Item specific)	A	.60	.00	.05	.04	.31
	B	.08	.01	.04	.33	.54
	C	.34	.21	.17	.11	.17
H_6 (Type specific)	A	.48	.01	.08	.21	.23
	B	.08	.00	.06	.32	.53
	C	.25	.15	.24	.30	.07
H_7 (Latent distance)	A	.58	.00	.05	.06	.32
	B	.08	.01	.04	.33	.54
	C	.34	.21	.12	.13	.18

TABLE 8

Threshold Points and Latent Class Scores $\hat{x}_t$ for the Latent Distance Model, Assuming Normality

Data set	Threshold/ Score	Latent class $t =$				
		1	2	3	4	5
A	Threshold $\hat{c}_t$	.20	.20	.32	.47	—
	Score $\hat{x}_t$	−.67	.20	.26	.39	1.12
B	Threshold $\hat{c}_t$	−1.41	−1.36	−1.15	−.10	—
	Score $\hat{x}_t$	−1.86	−1.39	−1.26	−.57	.73
C	Threshold $\hat{c}_t$	−.42	.15	.47	.90	—
	Score $\hat{x}_t$	−1.08	−.13	.30	.67	1.45

can be used to assign scores to the uth cell or response pattern in the table [see (10)]. The rationale for this expression is that the uth cell will in general be a mixture of the T latent classes; s_u is the average of the relevant X scores for this cell. To save space, these scores will not be presented here.

4.4. REMARKS

1. Scores obtained from (13) will be different from "scores" obtained by summing the items [i.e., $s_u^* = \sum_j k_j$, where $u = (k_1, \ldots, k_4)$].

2. Scores are available for all cells, or for all response patterns, or for all individuals. That is, the scaling does not ignore response patterns that are not scale types.

3. Distributional assumptions other than a normality assumption can of course be used. The logistic is a simple alternative to the normal and is virtually indistinguishable from it. The log-normal would be useful in cases where the underlying distribution is thought to be skewed. Many other possibilities could be considered.

4. When latent class probabilities are small, we see from Tables 7 and 8 that there is not much power to discriminate among individuals. For example, if $\pi_t^X = \varepsilon$, the scores $x_{t'-1}$ and $x_{t'}$ will be virtually indistinguishable [$x_{t'} = x_{t'-1} + o(\varepsilon)$]. In latent class analysis, it is usually the case that latent classes with small probabilities can be ignored, or the number of latent classes originally posited will be reduced. This practice

is justified by the scoring methods presented here as well as by the criterion of parsimony.

5. Many alternative scaling models are available. Goodman (1975) and Dayton and Macready (1980) consider models with *intrinsically unscalable respondents*. Their models incorporate assumptions about population heterogeneity with respect to scalability of the items. The general approach presented here can be modified to deal with these alternative formulations.

6. There are three key ingredients to the approach just discussed: (a) an *a priori* ordering of items, producing the scale types used to define the models; (b) specification of the relevant response-error model (restrictions on the α_{jt}); and (c) the distributional assumption for X^*, the implicit latent variable that we categorize to produce X. The approach considered next avoids assumptions (a) and (c).

7. The suggestion here is that the latent class model—the response-error model—can be defended without reference to a distributional assumption. We recommend using a distributional assumption only *after* the model has been tested and the latent class probabilities π_t^X have been estimated. The latent class model can be defended without reference to the distributional assumption. Conventional methods for diagnosing and/or modifying latent class models can be used prior to the actual scaling. Of course, we would obtain more efficient estimates of the x_t and the s_u if we incorporated the *correct* distributional assumption *a priori*.

5. LATENT-ASSOCIATION MODELS

5.1. MOTIVATION

The latent class scores x_t used in Sections 3 and 4 were obtained from a distributional assumption. The *ordering* of latent classes is unambiguous only when $T = 2$; when $T > 2$ restrictions must be applied. In Section 4 these restrictions were obtained from the item ordering as in Guttman's approach. The item ordering should ideally be gotten from a source outside the data being analyzed. Sometimes this will be difficult to do. The models considered in this section avoid these problems. They arise by extending the association models of Goodman (1984) and Clogg (1982) to the case where a latent categorical variable is considered as a grouping. Both the conditional association models and the partial association models in Clogg (1982) can be used to motivate results here.

5.2. The General Model—Dichotomous Items

The latent class model has been expressed earlier in terms of the π parameters [see (8)]. An equivalent way to represent the model is

$$\log[\pi(k_1, \ldots, k_J, x_t)] = \lambda + \lambda_{x(t)} + \sum_{j=1}^{J} \lambda_{j(k_j)} + \sum_{j=1}^{J} \lambda_{j(k_j), x(t)} \quad (14)$$

(See Haberman, 1979, Chapter 10.) The λ values in this expression can be determined from the π values in the earlier formulation. For example, when $T = 2$, $\lambda_{j(1), x(1)} = \frac{1}{4}\phi_{jx}$, where ϕ_{jx} is given by (9). The model in (14) leads to a different class of restricted models, however, because the parameters that we would restrict using (14) as a guide are functions of the parameters used earlier. For example, the model with the restrictions $\lambda_{1(1), x(1)} = \cdots = \lambda_{J(1), x(1)}$ can be considered easily with the formulation in (14), but it is very difficult to consider this model using the formulation in terms of π parameters.

We next consider latent class models with $T \geq 2$ latent classes. These classes will have *fixed* scores specified *a priori*, such as $x_t = t$ [or $x_t = t - (T + 1)/2$, for scores centered at zero]. The general model we wish to consider is the following:

$$\log[\pi(k_1, \ldots, k_J, x_t)] = \lambda + \lambda_{x(t)} + \sum_{j=1}^{J} \lambda_{j(k_j)} + \sum_{j=1}^{J} x_t \lambda_{j(k_j), x} \quad (15)$$

When $T = 2$, (15) is the conventional unrestricted two-class latent structure. The parameters in (14) have been restricted in (15) as $\lambda_{j(k_j), x(t)} = \lambda_{j(k_j), x}$; when $T \geq 3$ the model in (15) will differ substantially from conventional latent class models. The character of these restrictions is explored next.

Consider the association between latent variable X and item Y_1. When $T \geq 3$, the quantities

$$\phi_{1x(t)} = \log\left[\frac{\pi(1, k_2, \ldots, k_J, x_t)\pi(2, k_2, \ldots, k_J, x_{t+1})}{\pi(1, k_2, \ldots, k_J, x_{t+1})\pi(2, k_2, \ldots, k_J, x_t)}\right], \quad (16)$$
$$t = 1, \ldots, T - 1$$

can be used to describe this association. We see that $\phi_{1x(t)}$ is the logarithm of the odds ratio between items Y_1 and X when X takes on values t or $t + 1$, and when the other items (i.e., $Y_2, \ldots, Y_J$) are at specified levels. Using (16) and (15) we obtain

$$\phi_{1x(t)} = (x_t - x_{t+1})(2\lambda_{1(1)x}) \tag{17}$$

because $\lambda_{1(2)x} = -\lambda_{1(1)x}$. If $x_t - x_{t+1} = \delta$ for all t (equal spacing), then $\phi_{1x(t)}$ does not depend on t, or $\phi_{1x(t)} = \phi_{1x} = 2\delta\lambda_{1(1)x}$. Under equidistant spacing of X, we thus obtain for the jth item

$$\phi_{jx(t)} = \phi_{jx}, \qquad t = 1, \ldots, T - 1 \tag{18}$$

or one parameter is used to describe the entire relationship between X and item Y_j. Equation (18) says that the association between X and Y_j is uniform across levels t of X. This model is analogous to Haberman's (1974) model of linear-by-linear interaction; it is also analogous to Goodman's (1979) uniform association model; it is closely related as well to Clogg's (1982) models of uniform partial association. (See these references for details.)

Model (15) says that a T-class latent structure exists, with specified (e.g., equidistant) x category scores. The scores are used to restrict the X-by-item associations. This model can also be used as a measurement model. First note that the x_t values are given. Second, the score s_u for cell u in the original table is still obtained from (13). (To estimate s_u, we replace $P(X = t \mid u)$ by $\hat{P}(X = t \mid u)$.) Third, the item ordering is determined from the $\hat{\phi}_{jx}$ values, or an *a priori* ordering of items is not required.

5.3. RESULTS

We first consider the case where $T = 2$. The model in (14) is equivalent in this case to the model in (15); both are simple two-class latent structures if no restrictions on the parameter values are applied. We call this model H_8, although it is equivalent to model H_2 discussed in Section 3. When $T = 2$, scores are not relevant in the estimation. For contingency data set B, we found earlier that the two-class model had a boundary solution (see Table 3). When the alternative form of this model is considered here, the parameter estimates do not exist (specifically, $\hat{\lambda}_{1(1)x(1)} = \infty$). Because of this problem we consider only data sets A and C here.

Table 9 gives the L^2 values for H_8 and for some other models. Table 10 gives parameter estimates for some of the models in Table 9. Model H_8 ($= H_2$) of course gives the same L^2 values as were obtained earlier with H_2 (cf. Tables 3 and 9), but the parameter values reported in Table 10 are somewhat different from the parameter values in Table 4. Note

TABLE 9
Chi-Squared Values for Some Latent Association Models Applied to
Contingency Data Sets A and C in Table 1

Model	Description	Degrees of freedom	Likelihood-ratio-chi-square data set	
			A	C
$H_8(=H_2)$	Two-class model	6	67.29	14.29
H_9	H_8 with $\phi_{jx} = \phi_{.x}$	9	96.19	38.71
H_{10}	Restricted three-class model, fixed scores $[\lambda_{x(2)} = 0]$	6	51.91	11.57
H'_{10}	Restricted three-class model, fixed scores	5	48.03	—
H''_{10}	Restricted three-class model, fixed scores $[\lambda_{1(1)x} = \infty]$	6	17.72	—
H_{11}	$H_{10} + \phi_{jx} = \phi_{.x}$	9	80.19	35.34

TABLE 10
Estimated Parameter Values for Some Models in Table 9

Parameter	Data set						
	A				C		
	H_8	H_9	H_{10}	H'_{10}	H_8	H_9	H_{10}
$\hat{\lambda}_{x(1)}$	−.73	−.49	−.76	−1.80	.09	.43	.11
$\hat{\lambda}_{x(2)}$	—[a]	—[a]	.00[b]	3.36	—[a]	—[a]	.00[b]
$\hat{\lambda}_{1(1)}$	.72	.50	.72	.50	.55	.49	.54
$\hat{\lambda}_{2(1)}$	.25	.26	.25	.10	.19	.17	.19
$\hat{\lambda}_{3(1)}$	.13	.12	.13	.04	−.12	−.21	−.12
$\hat{\lambda}_{4(1)}$	.06	.05	.06	−.11	−.33	−.49	−.34
$\hat{\lambda}_{1(1),x}$	1.30	.95[c]	1.41	3.84	.65	.56[c]	.75
$\hat{\lambda}_{2(1),x}$	.83	.95[c]	.89	1.15	.75	.56[c]	.86
$\hat{\lambda}_{3(1),x}$	.90	.95[c]	.97	1.24	.45	.56[c]	.52
$\lambda_{4(1),x}$	.95	.95[c]	1.02	1.33	.35	.56[c]	.40

[a] Under H_8 and H_9, $\hat{\lambda}_{x(2)} = -\hat{\lambda}_{x(1)}$.
[b] A restriction.
[c] Under H_9, $\lambda_{j(1), x(1)} = \lambda_{j'(1), x(1)}$ for all items j and j'. Note that $\hat{\phi}_{jx} = 4\hat{\lambda}_{j(1), x(1)}$.

that $4\lambda_{j(1), x(1)} = 4\lambda_{j(1)x} = \phi_{jx}$, or the log-odds ratio between X and item Y_j is regarded as the key parameter here. The item ordering can be inferred from these values.

Next consider the hypothesis where

$$\lambda_{j(1), x(1)} = \lambda_{j'(1), x(1)}, \qquad \text{for all } j' \neq j \tag{19}$$

This condition states that $\phi_{jx} = \phi_{.x}$. If the items were test items, this condition is somewhat analogous to the hypothesis of "tau equivalent" tests (Lord, 1980). The two-class model with these constraints imposed is called H_9 in Table 9; the value of the difference, $L^2(H_9) - L^2(H_8)$, can be used as a test statistic for the condition in (19). We obtain differences of 28.90 for data set A and 24.42 for data set B, both indicating that the items are not equivalent.

We next consider model (15) with $T = 3$. Model H_{10} is the model where equidistant scores $(x_t = t)$ are used. This model was further restricted by imposing the condition $\lambda_{x(2)} = 0$; note that H_{10} has the same df as H_8. Parameter values appear in Table 10. Model H'_{10} is the model with the restriction $\lambda_{x(2)} = 0$ relaxed. We had difficulty in calculating the solution for data set C. Model H''_{10} is the model where $\lambda_{1(1)x}$ is allowed to approach infinity; this model was only estimated for Table 1, data set A. Model H_{11} is the model obtained from H_{10} by imposing the condition in (19). The difference between $L^2(H_{11})$ and $L^2(H_{10})$ can be used to test the condition.

Haberman's (1979, Appendix) program LAT was used to estimate these models. This program is based on the scoring algorithm, and good initial estimates must be supplied. MLLSA was used to determine these initial values, although we could not obtain satisfactory initial values for data set C for some models. We note that H'_{10} has unstable estimates for data set A (standard errors are available). This appears to arise from the fact that the latent variable X (with equally spaced scores) is closely associated with item Y_1 (see Table 10).

Scores for response patterns can be easily obtained from these models using the approach presented earlier.

5.4. Some Generalizations

The latent-association models are new, to our knowledge, although some relevant discussion (with an application) appears in Clogg and Shockey (1987). The point of these models is that the X variable is structured by (a) assuming scores x_t for the latent classes and (b) imposing a "linear-by-linear" interaction structure for the X-by-item associations. Some possible extensions are as follows: Suppose that the observed items are discrete-ordinal (e.g., Likert-type response items). If $\lambda_{j(k_j), x}$ denotes the value of the $X - Y_j$ interaction when $X_j = k_j$, model (15) still applies. If the levels of Y_j are equally spaced, the restriction, $\lambda_{j(k_j), x} - \lambda_{j(k_j+1), x} = \lambda_{jx^*}$ for $k_j = 1, \ldots, K_j - 1$, renders the $X - Y_j$ association *uniform* across levels of X and Y_j. When this restriction is not imposed, the values $\lambda_{j(k_j), x} - \lambda_{j(k_j+1), x} = \delta_{j(k_j)}$ can be used to infer the distances between levels of Y_j. [See, e.g., Clogg, 1982a; Goodman, 1984.] In principle, modeling strategies would attempt to refit models subject to suitable restrictions and varying the number of latent classes (T).

The primary difficulty in applying these models at the moment is computation. "Full information" algorithms like scoring or Newton–Raphson methods seem too sensitive to initial values. The EM algorithm employed in MLLSA is a possible alternative, although this has not been attempted at the present time.

6. SCALED LATENT CLASS MODELS

We next consider models where the scores for latent classes are parameters to be estimated. We let μ_t denote the score for the tth latent class. Before considering some general models, Rasch's (1960/1980, 1966) model will be discussed.

6.1. RASCH'S MODEL—DICHOTOMOUS ITEMS

For the moment assume that there are $T = n$ latent "classes," where n is the sample size. Let μ_i, $i = 1, \ldots, n$, denote the value of X for the ith individual (or for the ith "latent class"). For simplicity, let $Y_j = 1$ or 0 (instead of $Y_j = 1$ or 2); responses of "2" are recoded to "0". We add subscript i to denote the subject, so Y_{ij} refers to the response of the ith individual on the jth item, y_{ij} to the observed response. Let $\pi_{j|i}$ denote the conditional probability $P(Y_{ij} = 1 \mid X = \mu_i)$; this of course is the same as $P(Y_{ij} = 1)$.

Rasch's model says that

$$\pi_{j|i} = (e^{\mu_i - \alpha_j})/\Delta_{ij} \tag{20}$$

where

$$\Delta_{ij} = 1 + e^{\mu_i - \alpha_j} \tag{21}$$

is a normalizing constant ensuring that $\pi_{j(1)} + (1 - \pi_{j|i}) = 1$. The likelihood function simplifies to

$$L(\mu, \alpha \mid y) = D \exp\left(\sum_i \mu_i y_{i+} - \sum_j \alpha_j y_{+j}\right) \tag{22}$$

where $D^{-1} = \prod_i \prod_j \Delta_{ij}$ (see Andersen, 1980). The y_{i+} (number of "1" responses for the ith individual) are sufficient statistics corresponding to the μ_i, and the y_{+j} are the sufficient statistics corresponding to the α_j. (Note that y_{+j} and $n - y_{+j}$ give the marginal distribution of the jth item, so with n fixed the marginals are determined from the set of y_{+j} values.)

The μ_i are nuisance parameters, and inference based on (22) is complicated by this fact. A conditional likelihood argument (Andersen, 1980) is typically used when the number J of items is not large. The approach is to condition on the observed y_{i+} values; the conditional likelihood that results (which does not contain the μ_i values) can be used to make inferences about the item parameters α_j. Of course, different arguments are required in order to obtain inferences (e.g., estimates) about the ability parameters μ_i.

Note that the n y_{i+} values observed for J items actually take on only $J + 1$ values $0, 1, \ldots, J$. Let n_r, $r = 0, \ldots, J$, denote the number of subjects with *scores* $y_{i+} = r$, with $\sum_r n_r = n$. Let R refer to the score or *score group* variable. By Tjur (1982), Duncan (1984), or Kelderman (1984), the conditional form of Rasch's model can be represented as

$$\log(F_u) = \lambda + \lambda_{R(r)} + \sum_j \lambda_{j(k_j)} \tag{23}$$

where u is a cell in the contingency table, $u = \{(k_1, \ldots, k_J)\}$, $r = \sum_j k_j$, $\lambda_{R(r)}$ is the main effect of r, and $\lambda_{j(k_j)}$ is the main effect of the jth item. The usual constraints are assumed: $\sum_r \lambda_{R(r)} = 0$, $\lambda_{j(1)} + \lambda_{j(2)} = 0$. Even with these constraints, however, the model is still overparametrized, and we can remove the redundancy by setting $\lambda_{R(J)} = 0$. The $\lambda_{R(r)}$ values ensure that the marginal distribution of R, $(n_0, n_1, \ldots, n_J)$, will be fitted when maximum likelihood estimation is used. If we condition on the observed n_r values, the $\lambda_{R(r)}$ should not be regarded as parameters. As we will see below, an interesting class of models arises when we regard the $\lambda_{R(r)}$ as parameters that might be restricted in some fashion. The item parameters α_j can be retrieved from the $\lambda_{j(k_j)}$ values taking account of the other constraints.

The log-linear model in (23) says that the items are independent given variable R. Since any independence or conditional independence model for contingency tables can be represented as a latent class model,

TABLE 11
Chi-Squared Values for Rasch's Model and Some Related Models

Model	Description	Degrees of freedom	Likelihood-ratio chi-square		
			A	B	C
M_1	Rasch	8	47.15	7.31	29.39
M_2	Rasch + $R_L \times Y_j$	5	16.76	3.43	8.98
M_3	Rasch with $R_{\text{cubic}} = 0$	9	50.31	8.72	30.50
M_4	$M_3 + R_L \times Y_j$	6	20.94	6.09	11.29

the conditional form of Rasch's model is a latent class model. The precise relationship is that $X = R$, $\pi_r^X = n_r/n$ (conditionally fixed), and the items $Y_1, \ldots, Y_J$ are independent given X (or R). Note that the conditional form of Rasch's model is a T-class latent structure, with $T = J + 1$. This means that the individual-level heterogeneity allowed in the form of Rasch's model that would be considered when there are not many items is no different from that allowed in any $(J + 1)$-class latent structure.

6.2. THE RASCH MODEL APPLIED TO THE DATA IN TABLE 1

When Rasch's model is represented as the log-linear model for contingency tables [see (23)], it can be estimated and tested using a variety of different programs (MULTIQUAL, FREQ, SPSS^x, GLIM, SAS, etc.). Table 11 gives the L^2 values for Rasch's model, called M_1. We see that M_1 fits the data well for data set B, but it does not fit well for the other tables. Parameter values for M_1 appear in Table 12. Note that only estimates for the $\lambda_{j(1)}$, $j = 1, \ldots, 4$, are included, since the $\lambda_{R(r)}$ are not

TABLE 12
Parameter Values for Some of the Models in Table 11

Model	Data set	Parameter						
		$\lambda_{1(1)}$	$\lambda_{2(1)}$	$\lambda_{3(1)}$	$\lambda_{4(1)}$	$\lambda_{1(1)R_L}$	$\lambda_{2(1)R_L}$	$\lambda_{3(1)R_L}$
M_1	A	.68	.93	1.07	1.14	—	—	—
	B	.24	1.81	2.92	3.13	—	—	—
	C	.04	.36	.74	1.03	—	—	—
M_2	A	1.13	.59	.86	1.12	−.29	.15	.08
	B	−.10	1.94	2.57	2.12	−.21	−.61	−.29
	C	−.03	.45	.47	.45	−.22	−.29	−.11

regarded as parameters. The α_j parameters can be calculated from the values in Table 12, or they can be obtained from the fitted frequencies (see Duncan, 1984).

6.3. SOME MODELS RELATED TO RASCH'S MODEL

A property of Rasch's model is that the $X - Y_j$ association does not depend on the level of X; see (20) and (23). (Note that this property holds true in both the unconditional and the conditional form of the Rasch model.) This same property is actually used to define the item-specific response-error model (H_5 in Section 4) and the latent-association models H_{10} and H_{11} in Section 5. As before, this condition can be relaxed. A simple way to do this is to let $x_r = r$ denote the score for variable R (= variable X). Two-factor interaction parameters $\lambda_{j(k_j), R_L}$ would be added to the model of (23) using terms $x_r\lambda_{j(k_j), R_L}$, $j = 1, \ldots, J$. These refer to the extent to which the $X - Y_j$ association varies linearly over X levels; the model needs a constraint like $\lambda_{J(k_j), R_L} = 0$ to be meaningful. The model with $X - Y_j$ interactions of this kind is called M_2 in Table 11, and the parameter values appear in Table 12. Note the dramatic improvement in fit for data sets A and C. The parameter values $\lambda_{j(1), R_L}$, $j = 1, \ldots, 4$, for data set A are $-.29$, $.15$, $.08$, and $.00$, the latter of which is a restriction. To find the $X - Y_j$ association at any level r of X we take $\lambda_{j(1)} + (\lambda_{j(1), R_L})x_r$.

The scoring system $x_r = r$ is a necessary given for M_2, whereas it was not necessary for M_1. The scores x_r must be regarded as fixed. (Of course, some distributional assumption could be applied to the n_r values, which gives the distribution of R or X, to obtain different scores.) Note that there can be at most $J + 1$ different μ_r values since there are only $J + 1$ values of the sufficient statistics Y_{i+}.) In effect, the μ_r values are set at $x_r = r$ in M_2. This approach is very consistent with the "number right" scoring rule used in ability tests. Note that Rasch's model (M_1) does not depend on a scoring system at all, since the $\lambda_{R(r)}$ values are used only to preserve score group totals: no spacing assumptions for R levels are required in order to do this.

The preceding observations suggest further modifications of Rasch's model, which we consider next. (For related methods, see Kelderman, 1984.) Let us now suppose that the score group totals n_r are random, as they are in fact. It follows that the $\lambda_{R(r)}$ values can be treated as parameters, which might be further restricted. There are $J + 1$ $\lambda_{R(r)}$, J of which are nonredundant, and previously an additional restriction was imposed to identify parameters. Assume now that scores x_r are available for levels of R; equal-interval scores $x_r = r$ are the simplest, but any set of fixed scores can be used, although the model discussed next depends on the scoring system. The $\lambda_{R(r)}$ term in (23) can be replaced with

$$\lambda_{R_L} x_r + \lambda_{R_Q} x_r^2 + \cdots + \lambda_{R_*} x_r^J$$

where λ_{R_L} is the linear effect, λ_{R_Q} is the quadratic effect, and so forth. Under the scoring system $x_r = r$ we have $x_r = \sum_j k_j$ for cell u with score r, so the linear term cannot appear if all item main effects are in the model. If the other $J - 1$ terms are included, Rasch's model is obtained regardless of the scoring system, since all $J - 1$ degrees of freedom (or marginal constraints) are used. It is natural to consider models where one or more of the higher-degree terms are excluded. Model M_3 in Table 11 was obtained by setting λ_{R_L} and $\lambda_{R_{cort}}$ to zero ("Cort" for cortic term) and retaining the corresponding quadratic and cubic terms. As expected, the model obtained in this way is virtually indistinguishable from Rasch's model. The difference between $L^2(M_3)$ and $L^2(M_1)$ might serve as a check on the suitability of the scoring system, but of course other checks are possible.

Model M_4 is the model obtained from M_3 by allowing the $X-Y_j$ association (or $R-Y_j$ association) to vary linearly with X levels (or R levels), again assuming equal spacing of R levels. Note that M_3 and M_4 require scores x_r for the "latent" variable. Note also that the latent class proportions $\hat{\pi}_r$ are no longer equal to n_r/n—we do not condition on the score group totals when such modified models are considered.

6.4. Modified Rasch Models and Measurement

The models just discussed can be used directly to obtain measurements. The x_r values not only scale X but also the response patterns, since r is the score of the uth response pattern. Under the modified models M_3 and M_4, a probabilistic assignment is still unnecessary. Measurement is very straightforward with these models.

6.5. Another Modification of Rasch's Model: A Discrete Latent Trait Model

Let us assume that there are T latent classes, with $T < n$ (usually T will be much smaller than n). T does not depend on the sample size, but it might depend on the number of items. Let μ_t, $t = 1, \ldots, T$ denote the latent class scores, now regarded as parameters. Let $\pi_t^X = P(X = t)$ [or $P(X = \mu_t)$] as before. The model is

$$\pi_{j|t} = (e^{\mu_t - \alpha_j})/\Delta_{tj} \tag{24}$$

where

$$\Delta_{tj} = 1 + e^{\mu_t - \alpha_j} \tag{25}$$

This model has been considered by Formann (1984: p. 110); it is merely a discrete version of Rasch's model (when $T = n$ Rasch's model is obtained). We need to analyze the likelihood function in order to set up the analysis of this model.

Consider the joint probability of a given response pattern $(k_1, \ldots, k_J)$ for an individual in the tth latent class. (Recall the fact that the k_j are scored 0 or 1.) This can be written as

$$P(Y_{ij} = y_{ij}, j = 1, \ldots, J \mid X = t) = \prod_j \Delta_{tj}^{-1} \exp(\mu_t y_{ij} - \alpha_j y_{ij}) \tag{26}$$

using (24) and (25). The relevant unconditional probability for the ith individual is

$$
\begin{aligned}
P(Y_{ij} = y_{ij}, j = 1, \ldots, j) &= \sum_t \pi_t^X \prod_j \Delta_{tj}^{-1} \exp(\mu_t y_{ij} - \alpha_j y_{ij}) \\
&= \sum_t \pi_t^X \left[\exp\left(\mu_t y_{i+} - \sum_j \alpha_j y_{ij} \right) \right] \Big/ \prod_j \Delta_{tj}
\end{aligned}
\tag{27}
$$

Note that the axiom of local independence was used to obtain these expressions. With n independent observations, the likelihood function is

$$L(\pi_t^X, \alpha_j, \mu_t \mid y_{ij}) = \exp\left(-\sum \alpha_j y_{+j} \right) \prod_i \left\{ \sum \pi_t^X [\exp(\mu_t y_{it})] \Big/ \prod_j \Delta_{tj} \right\} \tag{28}$$

Thus, the item marginals y_{+j} are still sufficient statistics corresponding to the item parameters α_j, just as in Rasch's model. On the other hand, the sufficient statistics for μ_t are more complicated. (If we condition on the values of π_t^X, the kernel of the likelihood related to the μ_t values takes on a rather simple form, and this fact can be used to assist in the calculations.) At this point the problem has been reduced to a calculation procedure.

6.6. Results

The EM algorithm was applied to the model in (24) using the likelihood function developed in (26)–(28). Models with $T = 2, 3, 4,$ and 5 were considered for each data set; these models are called M_5, M_6, M_7,

TABLE 13
Chi-Squared Values for the Scaled Latent Class Model, or Discrete Latent
Trait Model

			Likelihood-ratio chi-square		
Model	Description	Degrees of freedom	A	B	C
M_5	$T = 2$ classes	9	96.19	56.13	38.71
M_6	$T = 3$ classes	8	47.15	7.31	29.39
M_7	$T = 4$ classes	8	47.15	7.31	29.39
M_8 $(= M_1)$	$T = 5$ classes [= Rasch]	8	47.15	7.31	29.39

and M_8, respectively. Chi-squared values appear in Table 13. The two-class model (M_5) adds two parameters to the independence model (H_1): one nonredundant latent class proportion and one nonredundant score parameter. (It can be verified that the independence model arises by setting $T = 1$ and positing a constant trait value, say μ.) Thus, model M_5 has $11 - 2 = 9$ degrees of freedom. It might be supposed that model M_6, the three-class model, would have $9 - 2 = 7$ degrees of freedom, but actually the parameters in M_6 are not all identifiable. The item parameters are identifiable, but the latent class proportions and the score parameters have an element of indeterminancy. Model M_6 actually has eight degrees of freedom, which is the same as the degrees of freedom for Rasch's model (M_1). As can be seen from the chi-squared values in Table 13, models M_6–M_8 are equivalent to each other, and each model gives the same estimated expected frequencies (and the same chi-squared values) as Rasch's model. The reasons for this equivalence and further comparisons among models are being taken up in a separate report. The scaled-latent-class model turns out to be equivalent to Rasch's model when the number T of latent classes is greater than or equal to $(J + 1)/2$, where J is the number of items. These results do not appear to be available in the literature.

An interesting feature of the models is that at most $T = 3$ latent classes are necessary to describe the interindividual intrinsic variability in the latent trait for four items. For dataset B, for example, with $L^2(M_6) = 7.31$ on eight degrees of freedom, it is not necessary to posit more than three distinct values of the latent trait. With Rasch's model, on the other hand, the latent trait values ostensibly take on n different values; with the conditional-likelihood representation of Rasch's model, it is commonly supposed that at least five latent classes are employed.

Table 14 gives the parameter estimates for model M_6, the three-class

TABLE 14
Parameter Values for Model M_6 in Table 13

Parameter	Data set		
	A	B	C
Latent class proportions			
π_1^X	.52	.09	.43
π_2^X	.32	.26	.49
π_3^X	.16	.66	.09
Latent class scores[a]			
μ_1	2.43	2.67	.61
μ_2	−.81	−4.08	−1.04
μ_3	−4.54	−6.98	−3.74
Item difficulties			
α_1	−.93	−5.80	−1.97
α_2	−.44	−2.65	−1.32
α_3	−.15	−.42	−.57
α_4	.00[b]	.00[b]	.00[b]

[a] Scores not identified; ratio of distances $(\hat{\mu}_1 - \hat{\mu}_2)/(\hat{\mu}_2 - \hat{\mu}_3)$ is identified.
[b] $\alpha_4 = 0$ is an identifying restriction.

model. The item difficulties α_j were identified by setting $\alpha_4 = 0$, a constraint that is familiar from analysis of the corresponding Rasch model. The score parameters provided in Table 14 are not identified, but the ratio of distances, $(\hat{\mu}_1 - \hat{\mu}_2)/(\hat{\mu}_2 - \hat{\mu}_3)$ is identified and hence invariant under any unit-point and zero-point restriction on these scores. These scores could be used to scale the response patterns and hence produce the measurements for the subjects using formula (13). Parameter values for the other models are similar and so will not be reported.

7. CONCLUSION

Scores for latent classes must be considered in order to reconcile latent class models and other latent structure models that are based on continuous or quantitative latent variables or latent traits. In this chapter several ways of incorporating scores were introduced. Even with the simple latent class model we can create meaningful methods for measurement, although the model itself does not incorporate those scores on an *a priori* basis. With the so-called scaling models that stem from various restricted latent class models, a clearer relationship between the discrete and the continuous theories can be seen. Finally, the latent-association models and the models motivated by Rasch's model offer ostensibly different ways in which to relate the different theories. I believe that further work in the direction outlined here will lead to an even greater reconciliation between the different theories of measurement. The latent class model provides a sensible framework for carrying out these next steps.

8. REFERENCES

Aitkin, M., Anderson, D., & Hinde, J. (1981). Statistical modeling of data on teaching styles. *Journal of the Royal Statistical Society, Series A, 144,* 419–461.

Andersen, E. B. (1980). *Discrete statistical models with social science applications.* Amsterdam: North-Holland.

Bartholomew, D. J. (1980). Scaling binary data using a factor model. *Journal of the Royal Statistical Society, Series B, 42,* 293–321.

Becker, M. P., & Clogg, C. C. (1986). A note on approximating correlations from odds ratios. Unpublished manuscript.

Bentler, P. M. (1983). Some contributions to efficient statistics in structural models: Specification and estimation of moment structures. *Psychometrika, 48,* 493–517.

Bishop, Y. M. M., Fienberg, S. E., & Holland, P. W. (1975). *Discrete multivariate analysis: Theory and Practice.* Cambridge, MA: MIT Press.

Cliff, N. (1977). A theory of consistency of ordering generalizable to tailored testing. *Psychometrika, 42,* 375–399.

Clogg, C. C. (1979). Some latent structure models for the analysis of Likert-type data. *Social Science Research, 8,* 287–301.

Clogg, C. C. (1981a). Latent structure models of mobility. *American Journal of Sociology, 86,* 836–868.

Clogg, C. C. (1981b). New developments in latent structure analysis. In D. M. Jackson & E. F. Borgatta (Eds.), *Factor analysis and measurement in sociological research* (pp. 214–280). Beverly Hills, California: Sage.

Clogg, C. C. (1982). Some models for the analysis of association in multi-way cross-classifications having ordered categories. *Journal of the American Statistical Association, 77,* 803–815.

Clogg, C. C. (1984). Some statistical models for analyzing why surveys disagree. In C. F. Turner & E. Margin (Eds.), *Surveying subjective phenomena* (Vol. 2, pp. 319–366). New York: Russell Sage Foundation.

Clogg, C. C., & Goodman, L. A. (1984). Latent structure analysis of a set of multidimensional contingency tables. *Journal of the American Statistical Association, 79,* 762–771.

Clogg, C. C., & Goodman, L. A. (1985). Simultaneous latent structure analysis in several groups. In N. B. Tuma (Ed.), *Sociological methodology 1985* (pp. 81–110). San Francisco: Jossey-Bass.

Clogg, C. C., & Goodman, L. A. (1986). On scaling models applied to data from several groups. *Psychometrika, 51,* 123–135.

Clogg, C. C., & Sawyer, D. O. (1981). A comparison of alternative models for analyzing the scalability of response patterns. In S. Leinhardt (Ed.), *Sociological methodology 1981* (pp. 240–280). San Francisco: Jossey-Bass.

Clogg, C. C., & Shockey, J. W. (1987). Multivariate analysis of discrete data. In J. R. Nesselroade & R. B. Cattell (Eds.), *Handbook of multivariate experimental psychology*. New York: Plenum.

Cochran, W. G. (1983). *Planning and analysis of observational studies*. New York: Wiley.

Dayton, C. M., & Macready, G. A. (1980). A scaling model with response errors and intrinsically unscalable respondents. *Psychometrika, 45,* 343–356.

Duncan, O. D. (1984). Rasch measurement: Further examples and discussion. In C. F. Turner & E. Martin (Eds.), *Surveying subjective phenomena* (Vol. 2, pp. 367–403). New York: Russell Sage Foundation.

Fienberg, S. E. (1980). *The analysis of cross-classified categorical data* (2nd ed.). Cambridge, MA: MIT Press.

Formann, A. K. (1984). *Die Latent-Class-Analyse*. Weinheim and Basel: Beltz Verlag.

Formann, A. K. (1985). Constrained latent class models: Theory and applications. *British Journal of Mathematical and Statistical Psychology, 38,* 87–111.

Goodman, L. A. (1974a). The analysis of systems of qualitative variables when some of the variables are unobservable: Part I. A modified latent structure approach. *American Journal of Sociology, 79,* 1179–1259.

Goodman, L. A. (1974b). Exploratory latent structure analysis using both identifiable and unidentifiable models. *Biometrika, 61,* 215–231.

Goodman, L. A. (1975). A new model for scaling response patterns: An application of the quasi-independence concept. *Journal of the American Statistical Association, 70,* 755–768.

Goodman, L. A. (1978). *Analyzing qualitative/categorical data*. Cambridge, MA: Abt Books.

Goodman, L. A. (1979). Simple models for the analysis of association in cross-classifications having ordered categories. *Journal of the American Statistical Association, 74,* 537–552.

Goodman, L. A. (1984). *The analysis of cross-classifications having ordered categories*. Cambridge, MA: Harvard University Press.

Guttman, L. (1950). The basis for scalogram analysis. In S. A. Stouffer, L. Guttman, E. A. Suchman, P. F. Lazarsfeld, S. A. Star, & J. A. Clausen (Eds.), *Measurement and prediction: Studies in social psychology in World War II* (Vol. 4, pp. 60–90). Princeton, NJ: Princeton University Press.

Haberman, S. J. (1974). Log-linear models for frequency tables with ordered classifications. *Biometrics, 30,* 589–600.

Haberman, S. J. (1979). *Analysis of qualitative data: Vol. 2. New developments*. New York: Academic Press.

Joreskog, K. G., & Sorbom, D. (1979). *Advances in factor analysis and structural equation models*. Cambridge, MA: Abt Books.

Kelderman, H. (1984). Loglinear Rasch model tests. *Psychometrika, 49,* 223–245.

Lawley, D. N., & Maxwell, A. E. (1967). *Factor analysis as a statistical method* (2nd ed.). New York: Elsevier.

Lazarsfeld, P. F., & Henry, N. W. (1968). *Latent structure analysis*. Boston: Houghton Mifflin.

Lord, F. M. (1980). *Applications of item response theory to practical testing problems*. Hillsdale, NJ: Lawrence Erlbaum.

Masters, G. N. (1985). A comparison of latent trait and latent class analyses of Likert-type data. *Psychometrika, 50,* 69–82.

McCutcheon, A. L. (1985). A latent class analysis of tolerance for nonconformity in the American public. *Public Opinion Quarterly, 49,* 474–488.

Mooijaart, A. (1982). Latent structure analysis for categorical variables. In K. G. Joreskog & H. Wold (Eds.), *Systems under indirect observation.* Amsterdam: North-Holland.

Muthen, B. (1979). A structural probit model with latent variables. *Journal of the American Statistical Association, 74,* 807–811.

Proctor, C. A. (1970). A probabilistic formulation and statistical analysis of Guttman scaling. *Psychometrika, 44,* 421–425.

Rasch, G. (1966). An individualistic approach to item analysis. In P. F. Lazarsfeld & N. W. Henry (Eds.), *Readings in mathematical social science.* Chicago: Science Research Associates.

Rasch, G. (1980). *Probabilistic models for some intelligence and attainment tests* (2nd ed.). Chicago: University of Chicago Press. (Original work published 1960).

Rindskopf, D. (1983). A general framework for using latent class analysis to test hierarchical and nonhierarchical learning models. *Psychometrika, 48,* 85–97.

Schuman, H., & Presser, S. (1984). The assessment of "no opinion" in attitude surveys. In K. F. Schuessler (Ed.), *Sociological methodology 1979* (pp. 241–275). San Francisco: Jossey-Bass.

Tjur, T. (1982). A connection between Rasch's item analysis model and a multiplicative Poisson model. *Scandinavian Journal of Statistics, 9,* 23–30.

CHAPTER NINE

Comparison of Latent Structure Models

ERLING B. ANDERSEN

1. INTRODUCTION

In this chapter we shall by means of two examples compare different models for the same data set. The purpose is to illustrate various models as regards their capability to explain essential structures in the given data. The first example is concerned with consumer complaint behavior. The data consist of the responses for 600 individuals on six questions each with two answer categories. In the second example we consider two dichotomous questions; both questions have been answered by 3398 schoolboys on two occasions.

In both these examples the data in their formal structure constitute contingency tables. Example 1 is a $2 \times 2 \times 2 \times 2 \times 2 \times 2$ table and Example 2 a $2 \times 2 \times 2 \times 2$ table. It is tempting to analyze such tables by a log-linear model, but as we shall see, such a model is not very suitable for the kind of analysis we are interested in. Instead we shall try with various forms of latent structure models. We shall not give references to textbooks as regards standard statistical techniques in the text of the paper. As general references we mention Andersen (1980); Bishop, Fienberg, and Holland (1975); and Goodman (1978).

ERLING B. ANDERSEN • Department of Statistics, University of Copenhagen, DK-1455 Copenhagen K, Denmark.

2. MODELS

As our main purpose is to illustrate the use of different models on the same data, we shall not go into details with the statistical analyses, but rather present the various models on comparable forms. The first task at hand is to decide on a common form for displaying the data. In both examples we have k dichotomous questions. We thus consider $n \times k$ random variables

$$
X_{ij} = \begin{cases} 1 & \text{if individual No. i answers yes on question } j \\ 0 & \text{if individual No. i answers no on question } j \end{cases}
$$

In latent structure analysis we usually work with a specific model for the probability distribution of X_{ij} for each i and j. In log-linear models for contingency tables it is assumed that the probability distribution of X_{ij} is independent of i.

The log-linear model is based on the concepts of interactions. Any two of the k variables $X_{i1}, \ldots, X_{ik}$ are linked by interaction parameters of various orders. The term *log-linear* indicates that the linking is linear in the logarithms of the probabilities. In order to conform with the notation we shall use later, we write the logarithm of the probabilities as

$$
\ln f(x_{i1}, \ldots, x_{ik}) = \tau_0 + \tau_1(x_{i1}) + \cdots + \tau_k(x_{ik}) + \tau_{12}(x_{i1}, x_{i2}) + \cdots \tag{1}
$$
$$
+ \tau_{123}(x_{i1}, x_{i2}, x_{i3}) + \cdots + \tau_{12\ldots k}(x_{i1}, \ldots, x_{ik})
$$

where

$$
f(x_{i1}, \ldots, x_{ik}) = P(X_{i1} = x_{i1}, \ldots, X_{ik} = x_{ik})
$$

A simple counting shows that (1) is overparametrized because there are only 2^k possible response patterns $x_{i1}, \ldots, x_{ik}$. This problem is resolved by introducing the constraints

$$
\sum_{x=0}^{1} \tau_j(x) = 0
$$

$$
\sum_{x_1=0}^{1} \sum_{x_2=0}^{1} \tau_{j_1 j_2}(x_1, x_2) = 0
$$

and so on. In fact these constraints mean that there is only one unconstrained τ_1, one inconstrained τ_{12}, and so on. We for example

easily derive

$$\tau_{123}(0, 0, 1) = -\tau_{123}(1, 0, 1) = \tau_{123}(1, 1, 1)$$

Thus $\tau_{123}(x_{i1}, x_{i2}, x_{i3})$ is equal to either τ_{123} or $-\tau_{123}$. It is even easy to see that the sign is $+$ if there is an even number of zeros and $-$ if the number of zeros is odd. With this observation we have, for example,

$$\ln f(0, 1, 0, 1) = \tau_0 - \tau_1 + \tau_2 - \tau_3 + \tau_4 - \tau_{12}$$
$$+ \tau_{13} - \tau_{14} - \tau_{23} + \tau_{24} - \tau_{34}$$
$$+ \tau_{123} - \tau_{124} + \tau_{134} - \tau_{234} + \tau_{1234}$$

In (1) $\tau_0, \tau_1, \ldots, \tau_k$ are called *main effects* while all other τ's are called *interactions*. There are two conflicting terminologies. Often τ_{12} is called a *first-order interaction*, τ_{123} a *second-order interaction* and so on. More and more often, however, the terminology *two-factor interactions* for $\tau_{12}, \tau_{13}, \ldots,$ *three-factor interactions* for $\tau_{123}, \tau_{124},$ and so on are used. Especially when we deal with interactions of high order, the latter terminology is more convenient.

The statistical analysis of contingency tables by log-linear models aims at reducing the number of interactions necessary to explain the data. The reduction is from high towards low interactions; that is, one tries to explain the data by as few interactions of low order as possible. When a final model has been accepted the model is interpreted through a Darroch–Lauritzen–Speed diagram or DLS diagram (cf. Darroch, Lauritzen, & Speed, 1980). Each variable is represented by a dot, and those dots, for which the corresponding variables are connected through at least one nonzero interaction, are connected by a line. If two variables are not connected at all in the DLS diagram, they are stochastically independent. If the connection between two variables can be broken by covering one or more variables, these variables are conditionally independent given the covered variables. Figure 1 shows a case of four variables, where variable 1 is independent of all other variables and variables 2 and 4 are independent given the value of variable 3. The

FIGURE 1. DLS diagram for a four-dimensional contingency table.

model could have been

$$\ln f(x_{i1}, x_{i2}, x_{i3}, x_{i4}) = \tau_0 + \tau_1(x_{i1}) + \tau_2(x_{i2}) + \tau_3(x_{i3})$$
$$+ \tau_4(x_{i4}) + \tau_{23}(x_{i2}, x_{i3}) + \tau_{34}(x_{i3}, x_{i4})$$

Conditional independence is written symbolically as

$$\text{var. } 2 \otimes \text{var. } 4 \mid \text{var. } 3$$

where the sign $\otimes$ reads "independent" and the sign $\mid$ reads "given." Ordinary independence is, in accordance with this, written as

$$\text{var. } 1 \otimes \text{var. } 2$$

or if a variable is independent of several variables

$$\text{var. } 1 \otimes \text{var. } 2 \text{ and var. } 3$$

Latent structure models are also based on the concept of conditional independence. But in contrast to the conditional independence considered above, we now look at independence given the value of an unobservable so-called latent variable. Suppose thus that the added variable θ is introduced in a DLS-like diagram as shown in Figure 2. The independence is then between all the original variables given θ.

Under a latent structure model all variables may well be stochastically dependent before the latent variable is introduced. Stochastic independence given θ obviously means that

$$P(X_{i1} = x_{i1}, \ldots, X_{ik} = x_{ik} \mid \theta) = P(X_{i1} = x_{i1} \mid \theta) \cdots P(X_{ik} = x_{ik} \mid \theta)$$

or

$$f(x_{i1}, \ldots, x_{ik} \mid \theta) = f(x_{i1} \mid \theta) \cdots f(x_{ik} \mid \theta) \tag{2}$$

FIGURE 2. DLS diagram for a latent structure model.

Because in the data considered x_{ij} only takes the values 0 and 1, we can write $f(x_{ij} \mid \theta)$ as

$$f(x_{ij} \mid \theta) = \begin{cases} p_j(\theta) & \text{for } x_{ij} = 1 \\ 1 - p_j(\theta) & \text{for } x_{ij} = 0 \end{cases}$$

Formula (2) then becomes

$$f(x_{i1}, \ldots, x_{ik} \mid \theta) = p_1(\theta)^{x_{i1}}[1 - p_1(\theta)]^{1-x_{i1}} \cdots p_k(\theta)^{x_{ik}}[1 - p_k(\theta)]^{1-x_{ik}} \tag{3}$$

In order to derive the marginal distribution of $X_{i1}, \ldots, X_{ik}$, we need the distribution of θ in the population, that is, a statistical model for the variation of θ if the n individuals are randomly drawn from a specific population. We first consider the case where it can be assumed that the actual value of θ for any given individual can only take m distinct values $\theta_1, \ldots, \theta_m$. Let φ_v be the probability that a randomly drawn individual has θ-value θ_v. The marginal distribution of $X_{i1}, \ldots, X_{ik}$ is then according to (3) given by

$$f(x_{i1}, \ldots, x_{ik}) = \sum_{v=1}^{m} p_{1v}^{x_{i1}}(1 - p_{1v})^{1-x_{i1}} \cdots p_{kv}^{x_{ik}}(1 - p_{kv})^{1-x_{ik}}\varphi_v \tag{4}$$

where $p_{jv} = p_j(\theta_v)$. The number of parameters in this model is $k \cdot m + m - 1$, because $\sum \varphi_v = 1$. The model given by (4) is called a *latent class model*. The somewhat unrealistic assumption of a limited number of possible values of θ is introduced in order to limit the number of parameters. If we drop this assumption we must put more structure on the probabilities $p_j(\theta)$ in (3), preferable in terms of a few parameters. In *latent trait theory* there are many suggestions for such parametrizations. We shall primarily consider the *Rasch model*, where

$$p_j(\theta) = \exp(\theta + \varepsilon_j)/[1 + \exp(\theta + \varepsilon_j)] \tag{5}$$

The parameters $\varepsilon_1, \ldots, \varepsilon_k$ are connected with the k variables, in contrast to θ, which is connected with an individual. They are often called *item parameters* in reference to the terminology "items" for questions in, for example, mental test theory. Owing to the additive form $\theta + \varepsilon_j$, the constraint

$$\sum_{j=1}^{k} \varepsilon_j = 0$$

is introduced. In order to derive the marginal distribution of $X_{i1}, \ldots, X_{ik}$, we once again need the population distribution of θ. If the variation of θ is continuous this distribution is described by the density function $\varphi(\theta)$ of θ for the variation of θ in the population. We can then write the marginal distribution of $X_{i1}, \ldots, X_{ik}$ as

$$f(x_{i1}, \ldots, x_{ik}) = \int f(x_{i1} \mid \theta) \cdots f(x_{ik} \mid \theta)\varphi(\theta)\, d\theta \tag{6}$$

Again, there are many possibilities for a specification of $\varphi(\theta)$ in terms of a few parameters. We choose for convenience

$$\varphi(\theta) = \sigma^{-1}\varphi_N((\theta - \mu)/\sigma) \tag{7}$$

where φ_N is the standard normal density. The latent variable is thus normally distributed in the population with mean value μ and variance σ^2. From (5), (6), and (7) we get

$$f(x_{i1}, \ldots, x_{ik}) = \exp\left(\sum_j \varepsilon_j x_{ij}\right) \int \exp(\theta x_{i.})$$
$$\times \prod_{j=1}^{k} [1 + \exp(\theta + \varepsilon_j)]^{-1}\sigma^{-1}\varphi_N((\theta - \mu)/\sigma)\, d\theta \tag{8}$$

where $x_{i.} = \sum_j x_{ij}$. Model (8) has $k - 1 + 2 = k + 1$ parameters.

The Rasch model possesses a special property, which allows us to establish the conditional independence given θ and consequently check the latent structure of the model without specifying the precise form of the population density $\varphi(\theta)$. From (3) and (5) we get

$$f(x_{i1}, \ldots, x_{ik} \mid \theta) = \exp\left(\sum_j \varepsilon_j x_{ij} + \theta x_{i.}\right) \bigg/ \prod_{j=1}^{k} [1 + \exp(\theta + \varepsilon_j)] \tag{9}$$

because $1 - p_j(\theta) = 1/[1 + \exp(\theta + \varepsilon_j)]$.

It follows that $X_{i.}$ is a sufficient statistic for θ in the conditional distribution given θ. Hence we can write (9) as

$$f(x_{i1}, \ldots, x_{ik} \mid x_{i.})f(x_{i.} \mid \theta)$$

where the first factor is independent of θ, but of course depends on the ε's. Whatever the form of $\varphi(\theta)$, we then get

$$f(x_{i1}, \ldots, x_{ik}) = f(x_{i1}, \ldots, x_{ik} \mid x_{i.})f(x_{i.}) \tag{10}$$

where

$$f(x_{i.}) = \int f(x_{i.} \mid \theta)\varphi(\theta)\, d\theta$$

The ε_j's can now be estimated based on the conditional likelihood given $x_{1.}, \ldots, x_{n.}$, and the marginal probabilities $f(x_{i.})$ for $x_{i.} = 0, \ldots, k$ can be estimated simply by the frequencies of individuals with $x_{i.} = 0, \ldots, k$ found in the sample. We may in fact introduce a *pseudo-latent-class model* as

$$f(x_{i1}, \ldots, x_{ik}) = f(x_{i1}, \ldots, x_{ik} \mid r)f_r \tag{11}$$

where $f_0, \ldots, f_k$ are unknown probabilities of obtaining the score $r = x_{i.}$.

In the second example we shall finally consider a model with a *two-dimensional latent variable* (θ_1, θ_2). The situation could, for example, be that the same k questions are asked of the same individuals at two consecutive points in time, or the situation could be that $k = k_1 + k_2$ questions are being asked, k_1 of which refer to one latent variable θ_1, while the remaining k_2 refer to a second latent variable θ_2. Because the two sets of questions in both cases are answered by the same individuals, it is most likely that θ_1 and θ_2 are correlated. We are thus led to a model where θ_1 and θ_2 are connected through a two-dimensional latent density $\varphi(\theta_1, \theta_2)$, which we for convenience assume to be a normal density, that is,

$$\varphi(\theta_1, \theta_2) = \varphi_N(\theta_1, \theta_2 \mid \mu_1, \sigma_1^2, \mu_2, \sigma_2^2, \rho)$$

where μ_1 and σ_1^2 are the mean and variance of θ_1, μ_2 and σ_2^2 are the mean and variance of θ_2, and ρ is the correlation coefficient between θ_1 and θ_2.

Let now $X_{i1}, \ldots, X_{ik_1}$ be the answers of individual i to the first set of k_1 questions and $Y_{i1}, \ldots, Y_{ik_2}$ be the answers of the same individual to the second set of k_2 questions, where $k_1 = k_2$ in the first case discussed above.

The joint probability of $X_{i1} = x_{i1}, \ldots, X_{ik_1} = x_{ik_1}$ and $Y_{i1} = y_{i1}, \ldots, Y_{ik_2} = y_{ik_2}$ given θ_1 and θ_2 can then be written as

$$f(x_{i1}, \ldots, x_{ik_1}, y_{i1}, \ldots, y_{ik_2} \mid \theta_1, \theta_2)$$
$$= \prod_{j_1=1}^{k_1} p_{j_1}^{(1)}(\theta_1)^{x_{ij_1}}[1 - p_{j_1}^{(1)}(\theta_1)]^{1-x_{ij_1}} \prod_{j_2=1}^{k_2} p_{j_2}^{(2)}(\theta_2)^{y_{ij_2}}[1 - p_{j_1}^{(2)}(\theta_2)]^{1-y_{ij_2}} \tag{12}$$

where

$$p_{j_1}^{(1)}(\theta_1) = P(X_{ij_1} = 1 \mid \theta_1), \qquad j_1 = 1, \ldots, k_1 \tag{13}$$

and

$$p_{j_2}^{(2)}(\theta_2) = P(Y_{ij_2} = 1 \mid \theta_2), \qquad j_2 = 1, \ldots, k_2 \tag{14}$$

The marginal probability of the x's and y's is then given by

$$f(x_{i1}, \ldots, x_{ik_1}, y_{i1}, \ldots, y_{ik_2})$$
$$= \iint f(x_{i1}, \ldots, x_{ik_1}, y_{i1}, \ldots, y_{ik_2} \mid \theta_1, \theta_2)\varphi(\theta_1, \theta_2)\, d\theta_1\, d\theta_2 \tag{15}$$

where the conditional probability is given by (12).

As for the one-dimensional case, we can assume various models. If both θ_1 and θ_2 can only take a few possible values we have a *latent class model*.

Let the possible values of θ_1 be $\theta_{11}, \ldots, \theta_{1m_1}$ and the possible values of θ_2 be $\theta_{21}, \ldots, \theta_{2m_2}$ and let $\varphi_{v_1v_2}$ be the probability that $(\theta_1, \theta_2) = (\theta_{1v_1}, \theta_{2v_2})$ for a randomly drawn individual. The latent class model is then

$$f(x_{i1}, \ldots, x_{ik_1}, y_{i1}, \ldots, y_{ik_2})$$
$$= \sum_{v_1=1}^{m_1} \sum_{v_2=1}^{m_2} \prod_{j_1=1}^{k_1} (p_{j_1v_1}^{(1)})^{x_{ij_1}}(1 - p_{j_1v_1}^{(1)})^{1-x_{ij_1}} \tag{16}$$
$$\times \prod_{j_2=1}^{k_2} (p_{j_2v_2}^{(2)})^{y_{ij_2}}(1 - p_{j_2v_2}^{(2)})^{1-y_{ij_2}}\varphi_{v_1v_2}$$

where

$$p_{j_1v_1}^{(1)} = P(X_{ij_1} = 1 \mid \theta_{1v_1})$$

and

$$p_{j_2v_2}^{(2)} = P(Y_{ij_2} = 1 \mid \theta_{2v_2})$$

In the latent class model (15) we have $k_1 \cdot m_1 + k_2 \cdot m_2 + m_1 \cdot m_2 - 1$ parameters, because we have

$$\sum_{v_1=1}^{m_1} \sum_{v_2=1}^{m_2} \varphi_{v_1v_2} = 1$$

In the example below we shall consider the case $k_1 = k_2 = m_1 = m_2 = 2$, such that the number of parameters becomes 11.

If we assume that θ_1 and θ_2 both have continuous variations, we must parametrize the probabilities (13) and (14) in order to keep the number of parameters down. Again we shall use the *Rasch model* and assume that

$$p_{j_1}^{(1)}(\theta_1) = \exp(\theta_1 + \varepsilon_{1j_1})/[1 + \exp(\theta_1 + \varepsilon_{1j_1})] \tag{17}$$

and

$$p_{j_2}^{(2)}(\theta_2) = \exp(\theta_2 + \varepsilon_{2j_2})/[1 + \exp(\theta_2 + \varepsilon_{2j_2})] \tag{18}$$

where $\varepsilon_{11}, \ldots, \varepsilon_{1m_1}$ are the item parameters of the first m_1 items and $\varepsilon_{21}, \ldots, \varepsilon_{2k_2}$ the item parameters of the second set of items.

With these model specifications and a normal density for $\varphi(\theta_1, \theta_2)$, the model (15) has $k_1 - 1 + k_2 - 1 + 5$ parameters. In case $k_1 = k_2 = 2$, the model thus has seven parameters as compared with the 11 parameters for the latent class model. Under the Rasch model $X_{i.}$ and $Y_{i.}$ are sufficient statistics for θ_1 and θ_2 in the conditional model (12). Hence we can also here form a *pseudo-latent-class* model by grouping the individuals according to the values $(x_{i.}, y_{i.})$ of $(X_{i.}, Y_{i.})$. This gives the model

$$
\begin{aligned}
f(x_{i1}, &\ldots, x_{ik_1}, y_{i1}, \ldots, y_{ik_2}) \\
&= f(x_{i1}, \ldots, x_{ik_1} \mid x_{i.})f(y_{i1}, \ldots, y_{ik_2} \mid y_{i.})f(x_{i.}, y_{i.})
\end{aligned} \tag{19}
$$

where

$$f(x_{i.}, y_{i.}) = \iint f(x_{i.}, y_{i.} \mid \theta_1, \theta_2)\varphi(\theta_1, \theta_2)\, d\theta_1\, d\theta_2$$

In this model $\varepsilon_{11}, \ldots, \varepsilon_{1k_1}$ and $\varepsilon_{21}, \ldots, \varepsilon_{2k_2}$ can be estimated from the two first factors in (19) and we can estimate $f(x_{i.}, y_{i.})$ by the observed frequency of $(x_{i.}, y_{i.})$ in the sample. This latter estimation is independent of our choice of φ and hence the model can be checked without specifying the form of $\varphi(\theta_1, \theta_2)$. In this pseudo-latent-class model the number of parameters is $k_1 - 1 + k_2 - 1 + (k_1 + 1)(k_2 + 1) - 1$. In case $k_1 = k_2 = m_1 = m_2 = 2$, we have 10 parameters.

3. CHECKING THE MODEL

In the examples we leave out all details as regards the estimation of parameters, such that the emphasis is on comparison of models. Hence, it is most convenient to refer to the same basic test. In spite of the fact that there exist different goodness of fit tests for the different models, we shall thus check all models by the same standard χ^2 goodness-of-fit test, also when a different test is more appropriate. Since the basic data for the one-dimensional case are the response patterns $(x_{i1}, \ldots, x_{ik})$ we shall use the test statistic

$$q_1 = \sum_{x_{i1}=0}^{1} \cdots \sum_{x_{ik}=0}^{1} [o(x_{i1}, \ldots, x_{ik}) - e(x_{i1}, \ldots, x_{ik})]^2 / e(x_{i1}, \ldots, x_{ik}) \quad (20)$$

where $o(x_{i1}, \ldots, x_{ik})$ is the observed number of individuals in the sample with response patterns $(x_{i1}, \ldots, x_{ik})$ and

$$e(x_{i1}, \ldots, x_{ik}) = nf(x_{i1}, \ldots, x_{ik})$$

Because there are $2^k - 1$ possible response patterns, the degrees of freedom for the χ^2 distribution, which approximates the distribution of (19) when n is large, are

$$df = 2^k - 1 - g$$

where g is the number of parameters in the chosen model.

In the two-dimensional case, the test quantity is

$$g_2 = \sum_{x_{i1}=0}^{1} \cdots \sum_{x_{ik_1}=0}^{1} \sum_{y_{i1}=0}^{1} \cdots \sum_{y_{ik_2}=0}^{1}$$
$$\times [o(x_{i1}, \ldots, y_{ik_2}) - e(x_{i1}, \ldots, y_{ik_2})]^2 / e(x_{i1}, \ldots, y_{ik_2}) \quad (21)$$

where $o(x_{i1}, \ldots, y_{ik_2})$ is the observed number of individuals in the sample with response pattern $(x_{i1}, \ldots, x_{ik_1}, y_{i1}, \ldots, y_{ik_2})$ and

$$e(x_{i1}, \ldots, y_{ik_2}) = nf(x_{i1}, \ldots, x_{ik_1}, y_{i1}, \ldots, y_{ik_2})$$

4. EXAMPLES

4.1. EXAMPLE 1

The complaint data were introduced and first analyzed by Poulsen (1981). The present comparative analysis was first presented in Andersen

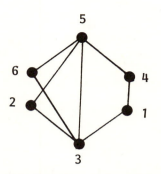

FIGURE 3. DLS diagram for complaint data.

(1982). The complaint data consist of six questions concerning consumer behavior. The 600 individuals were presented with six situations, where the purchased goods did not live up to the expectations. They were then asked whether or not they intended to complain to the shop in question. Many yes answers accordingly indicate a high degree of dissatisfaction. In Table 1, columns 1–6 show all response patterns $(x_{i1}, \ldots, x_{i6})$ that at least one individual has used. In column 7 of the table the observed numbers $o(x_{i1}, \ldots, x_{i6})$ are shown. The remaining columns are expected numbers under various models.

We start by analyzing the data by means of a log-linear model. It was found that a good fit was obtained with all six-factor, all five-factor, all four-factor, all three-factor interactions and in addition τ_{12}, τ_{15}, τ_{16}, τ_{24}, τ_{26}, τ_{34} and τ_{46} set equal to zero. This model gives the DLS diagram shown in Figure 3.

The interpretation of the model follows easily from the DLS diagram. The following conditional independencies thus characterize the data:

$$\text{var. 2} \otimes \text{var. 1 and var. 4} \mid \text{var. 3 and var. 5}$$
$$\text{var. 6} \otimes \text{var. 1 and var. 4} \mid \text{var. 3 and var. 5}$$
$$\text{var. 2} \otimes \text{var. 6} \mid \text{var. 3 and var. 5}$$

This model thus points to strong internal dependencies, where variable 3 and variable 5 seem to explain much of the variation in the data. The expected numbers under the fitted log-linear model are shown in column 8 of Table 1. The complaint data were analyzed by Poulsen (1981) using latent class analysis. The parameter estimates obtained are shown in Table 2. These parameters refer to model (4).

The expected numbers for this model are shown in column 9 of

TABLE 1
Observed and Expected Numbers under Four Different Models Fitted
to Danish Consumer Complaint Data

Response pattern (1–6)	Observed numbers (7)	Expected numbers			
		Log-linear (8)	Latent class (9)	Latent structure (10)	Pseudo-latent-class (11)
111111	127	115.4	129.1	118.6	127.0
111110	78	87.0	76.9	92.8	83.1
111101	16	18.2	16.5	33.7	30.2
111100	54	59.8	60.7	52.4	50.7
111011	36	40.9	32.7	24.9	22.3
111010	36	30.8	35.5	38.7	37.4
111001	14	15.9	12.5	14.1	13.6
111000	59	51.8	51.1	40.4	51.1
110111	8	6.4	7.4	8.0	7.2
110110	16	14.4	11.8	12.5	12.1
110101	4	1.8	4.8	4.5	4.4
110100	18	17.9	20.0	13.1	16.5
110011	1	2.2	3.1	3.4	3.2
110010	6	5.1	8.8	9.6	12.2
110000	18	15.5	17.0	18.2	15.9
101111	12	11.9	9.5	6.8	6.1
101110	7	9.0	10.4	10.6	10.3
101101	4	4.1	3.7	3.9	3.7
101100	18	13.6	15.0	11.1	14.0
101011	1	4.2	3.2	2.8	2.8
101010	6	3.2	6.9	8.2	10.3
101001	4	3.6	3.0	3.0	3.8
101000	9	11.8	12.7	15.4	13.5
100111	1	1.6	0.9	0.9	0.9
100110	4	3.7	2.6	2.6	3.3
100101	1	1.0	1.2	1.0	1.2
100100	10	10.2	5.0	5.0	4.3
100011	1	0.6	0.2	0.7	0.9
100010	3	1.3	2.1	3.7	3.2
100001	1	0.9	1.0	1.3	1.2
100000	3	8.8	4.2	12.6	5.1
011110	2	0.9	1.2	1.7	1.6
011101	1	0.2	0.6	0.6	0.6
011100	1	0.6	2.5	1.8	2.2
011011	1	2.2	0.2	0.5	0.4
011010	2	1.7	1.1	1.3	1.7
011001	3	0.9	0.5	0.5	0.6
010101	1	0.1	0.2	0.6	0.2
010100	1	0.9	0.8	0.8	0.7
010000	2	4.0	1.9	2.0	0.8
001010	1	0.2	0.4	0.5	0.4
001000	1	0.6	1.3	1.7	0.7
000010	1	0.3	1.2	0.4	0.2
000000	7	2.3	8.7	2.5	7.0

TABLE 2

Parameter Estimates of the Latent Class
Model (4) with $m = 3$ Fitted to the
Complaint Data

$\hat{p}_{jv}$	$v = 1$	2	3
$j = 1$	0.96	0.00	1.00
2	0.80	0.12	0.94
3	0.75	0.08	0.96
4	0.54	0.00	0.82
5	0.33	0.12	0.98
6	0.19	0.00	0.72
$\hat{\varphi}_v$	0.59	0.02	0.39

Table 1. We next analyze the data by the latent structure model (8). The parameter estimates for this model are shown in Table 3. The expected numbers under this model are shown in column 10 of Table 1. We finally consider the pseudo-latent-class model (11). The estimates of the score group frequencies $f_0, \ldots, f_6$ are shown in Table 4.

The expected numbers under the pseudo-latent-class model (11) are shown in column 11 of Table 1. The various models are compared through the goodness-of-fit test statistics (20) shown in Table 5.

It should be noted that the degrees of freedom in Table 5 are not $2^k - g$. Because of the large number of response patterns with small expected values, we have grouped the response patterns into 28 response

TABLE 3

Parameter Estimates of the
Latent Structure Model (8)

	$\hat{\varepsilon}_j$
$j = 1$	+2.47
2	+0.64
3	+0.48
4	−0.66
5	−0.96
6	−1.97
$\hat{\mu}$	1.408
$\hat{\sigma}^2$	1.405

TABLE 4
Score Group Frequencies
$f_0, \ldots, f_6$ under Model (11)
for the Complaint Data

	$\hat{f}_r$
$r = 0$	0.004
1	0.030
2	0.084
3	0.164
4	0.244
5	0.278
6	0.196

pattern groups. Hence

$$\text{df} = 28 - 1 - g$$

where g is the number of parameters in the model.

The only fit that is close to being satisfactory in Table 5 is the log-linear model. It is difficult, however, to translate the interpretation in terms of conditional independencies into useful new knowledge gained from the data analysis about consumer complaint behavior. The latent structure model, on the other hand, offers, if fitted to the data, an interpretation in terms of complaint behavior related to a latent complaint tendency. In addition the individuals can be categorized according to increasing complaint tendency, and the concrete answers to the six questions can be directly linked to the latent variable through the conditional probabilities $p_j(\theta)$ or p_{jv}, $j = 1, \ldots, 6$. Accordingly there are good reasons for scrutinizing the goodness of fit of the latent structure models more

TABLE 5
Comparison between the Goodness of Fit of the Various Models
Applied to the Complaint Data

Model	Formula	q	df	Significance[a] probability
Log-linear	(1)	24.2	13	0.031
Latent class	(4)	21.3	7^b	0.004
Latent structure	(8)	60.2	20	<0.001
Pseudo-latent-class	(11)	47.4	16	<0.001

[a] $P(Q \geq q)$, where $Q \sim \chi^2(\text{df})$.
[b] or 11 if the 0's and 1's of Table 2 are regarded as known values.

closely. In fact the fit is not that bad. An analysis of the single terms of the q-test quantities reveals that it is only for a few of the response patterns that the fit is really bad. Also by simply letting the eye scan the columns of Table 1, it is easy to see that the fit between observed and expected numbers is generally good. In this situation one would be tempted to accept the latent structure model as a tentative description of the data, maybe combined with a closer examination of the data corresponding to response patterns with significant deviations between observed and expected numbers.

Table 1 also requires a few remarks on the application of Pearson χ^2 tests. In this example we are faced with a contingency table with 64 cells and a sample of 600 individuals. As should be expected, there are accordingly few or no observations in several cells.

For the fitted models this means expected numbers close to zero. In this situation we have grouped the response patterns in order to make the χ^2 approximation valid. This grouping can be made in many different ways, and we are thus forced to introduce an element of arbitrariness into the data analysis. Unfortunately the choice of grouping may under unfortunate circumstances influence the conclusions to be drawn.

An even more serious problem is that standard statistical programs seldom check for this violation of the assumptions for the χ^2 approximation. In our first analysis of the data in Table 1, under a log-linear model, we thus got a computer output of $q = 72.2$ on 49 degrees of freedom. The significance probability of this result is 1.7%, although in Table 5 we found a significance probability of 3.1%. In the first analysis by Poulsen (1981) by means of a latent class model a q value of 41.5 was found on 43 degrees of freedom. This corresponds to a significance probability of 53.5% where we found 0.4%. This simple observation strongly supports the general rule for statistical practice of reexamining the original data many times during the data analysis.

The latent structure model and the latent class models are of course the same type of model. We should thus expect models (4) and (8) to be identical apart from the fact that θ in (4) only takes a limited number of values. We can illustrate this by comparing the parameters $p_j(\theta)$ given by (5) and p_{jv} given by

$$p_{jv} = p_j(\theta_v)$$

The problem is to locate θ_1, θ_2, and θ_3 on the θ scale. If for $v = 1, 2$, and 3 we solve the equations

$$p_{jv} = \exp(\theta_v + \varepsilon_j)/[1 + \exp(\theta_v + \varepsilon_j)] \tag{22}$$

TABLE 6

Comparison of Estimated Latent Class Probabilities p_{jv} and Calculated Rasch
Probabilities $p_j(\theta)$ for $\theta = 0.6$, -3.3, and 2.8

	p_{jv} $v = 1$	$p_j(\theta)$ $\theta = 0.6$	p_{jv} $v = 2$	$p_j(\theta)$ $\theta = -3.3$	p_{jv} $v = 3$	$p_j(\theta)$ $\theta = 2.8$
$j = 1$	0.96	0.96	0.00	0.30	1.00	0.99
2	0.80	0.78	0.12	0.07	0.94	0.97
3	0.75	0.75	0.08	0.06	0.96	0.96
4	0.54	0.49	0.00	0.02	0.82	0.89
5	0.33	0.41	0.12	0.01	0.98	0.86
6	0.19	0.20	0.00	0.01	0.72	0.70

for each j and calculate the average of the six values obtained, we get the
θ values

$$\theta_1 = 0.6, \qquad \theta_2 = -3.3, \qquad \theta_3 = 2.8$$

In Table 6 the values p_{jv} obtained from (22) with the three values above
inserted are shown together with p_{jv} in Table 2. The strong similarities
suggest that the latent class model is an approximation to the latent
structure model, where the latent variable is forced to take only three
possible values.

Another way to compare the two models is by constructing a table
showing the score $r = x_{i.}$, which is the sufficient statistic for θ under
model (8) with the assigned latent class for each individual under model
(4). The assigned latent class is the class for which the observed response
patterns has the highest probability. This comparison is carried out in
Table 7.

The boxes in Table 7 mark the agreement if scores 0 and 1 are

TABLE 7

Observed Number of Individuals for Each Score and Assigned
Latent Class for the Complaint Data

Assigned latent class	Score group						
	$r = 0$	1	2	3	4	5	6
$v = 1$	0	3	42	124	141	16	0
2	7	4	1	0	0	0	0
3	0	0	0	0	1	134	127

assigned to latent class 2, scores 2, 3, and 4 to latent class 1, and scores 5 and 6 to latent class 3.

4.2. EXAMPLE 2

We shall in this example reanalyze the famous Coleman data. The basic data and the reference to Coleman can be found in Goodman (1978, pp 417–418). There are two questions concerning attitude towards and membership in the leading crowd. These two questions are put at two different points in time to 3398 schoolboys. Because altogether four questions with two answer categories have been asked, there are 16 possible response patterns. The complete data set is shown in Table 8, columns 1–5.

The various models we shall consider are illustrated in Figure 4. Model I is a one-dimensional latent structure model of the form (8) with $k = 4$ questions. The estimated parameters are

$$(\hat{\varepsilon}_1, \ldots, \hat{\varepsilon}_4) = (-0.54, +0.35, -0.31, +0.50)$$

$$(\hat{\mu}, \hat{\sigma}^2) = (-0.15, 1.42)$$

TABLE 8
Observed and Expected Numbers for Five Different Models Fitted to the Coleman Data

Time 1		Time 2		Observed numbers	Expected numbers model				
Membership (1)	Attitude (2)	Membership (3)	Attitude (4)	(5)	I (6)	II (7)	IIIa (8)	IIIb (9)	IIIc (10)
1	1	1	1	458	417.4	281.9	454.9	455.6	452.2
1	1	1	0	140	99.4	108.3	144.1	157.8	146.3
1	1	0	1	110	225.5	226.9	109.1	148.4	119.9
1	1	0	0	49	105.3	135.5	48.9	64.5	49.4
1	0	1	1	171	116.6	133.6	172.1	157.8	174.5
1	0	1	0	182	54.4	74.4	179.8	179.7	177.5
1	0	0	1	56	123.4	155.9	58.3	64.5	58.8
1	0	0	0	87	110.7	135.0	85.8	94.1	74.4
0	1	1	1	184	282.1	288.5	188.6	148.4	174.8
0	1	1	0	75	131.7	160.6	68.8	64.5	71.9
0	1	0	1	531	298.8	336.7	530.4	530.7	531.4
0	1	0	0	281	267.9	291.3	283.2	310.6	281.8
0	0	1	1	85	154.5	190.1	82.1	64.5	85.6
0	0	1	0	97	138.5	153.3	101.5	94.1	108.3
0	0	0	1	338	314.1	321.3	337.5	310.6	336.1
0	0	0	0	554	557.6	404.7	552.9	552.2	555.1

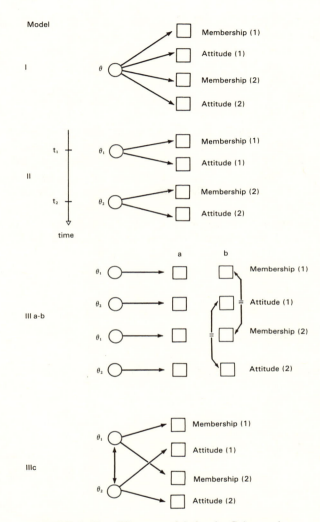

FIGURE 4. Five different models for the Coleman data.

The expected numbers under this model are shown in column 6 of Table 8. The fit is obviously not very good. Hence we shall abolish the idea of a one-dimensional latent variable. The same conclusion was drawn by Goodman (1978), who tried to fit a latent class model to the data. As an alternative he suggested a latent class model based on two latent variables, one that influences the response to the attitude questions and one that influences the response to the membership questions. We shall return to this model. But first we shall study Model II, which in a sense is the more obvious model. We have the same two questions, which are asked at two different points in time. Hence the natural assumption is that the opinions of the individuals change, while the conception of the questions remains the same. This would mean that the latent parameter changes from time point 1 to time point 2, but that the question characteristic does not change. Accordingly we should expect model (15) to fit the data with the conditional probability given by (12) and with $p_{j_1}^{(1)}(\theta_1)$ and $p_{j_2}^{(2)}(\theta_2)$ given by (17) and (18). This model is illustrated in Figure 4 as Model II. It is further to be expected that $\varepsilon_{1j} = \varepsilon_{2j}$ for $j = 1$, 2 and that θ_1 and θ_2 are to some extent correlated. Under the assumption of a two-dimensional normal density $\varphi(\theta_1, \theta_2)$, we found the parameter estimates

$$\hat{\varepsilon}_{11} = -0.385$$

$$\hat{\varepsilon}_{12} = +0.385$$

$$\hat{\varepsilon}_{21} = -0.370$$

$$\hat{\varepsilon}_{22} = +0.370$$

$$\hat{\mu}_1 = -0.213$$

$$\hat{\sigma}_1^2 = 0.482$$

$$\hat{\mu}_2 = -0.050$$

$$\hat{\sigma}_2^2 = 0.667$$

and

$$\hat{\rho} = 0.990$$

The fit of this model is, however, very bad, as can be seen directly from the expected numbers shown in Table 8, column 7. We, therefore, return to the suggestion by Goodman (1978). His model is illustrated in Figure 4 as Model IIIa. Models IIIb and IIIc in Figure 4 are also based on the assumption that the first latent variable influences the membership questions and the second latent variable influences the attitude questions. Model IIIa is a latent class model where each of the latent variables can take two values, such that we have four latent classes. For each latent variable we have two questions. Hence the model is (16) with $k_1 = k_2 = 2$ and $m_1 = m_2 = 2$. The estimated parameters, derived by Goodman, are shown in Table 9 with the notations of this paper. It should be noted

TABLE 9
Parameter Estimates under Models IIIa and IIIb

		Model			
		IIIa	IIIb	IIIa	IIIb
$\hat{p}^{(1)}_{j_1 v_1}$:		$v_1 = 1$		2	
	$j_1 = 1$	0.754	0.827	0.111	0.096
	2	0.910	0.827	0.076	0.096
$\hat{p}^{(2)}_{j_2 v_2}$:		$v_2 = 1$		2	
	$j_2 = 1$	0.806	0.821	0.267	0.287
	2	0.833	0.821	0.302	0.287
$\hat{\varphi}_{v_1 v_2}$:		$v_2 = 1$		2	
	$v_1 = 1$	0.271	0.272	0.129	0.130
	2	0.231	0.228	0.369	0.371

that under this model θ_1 influences the two membership questions, which in Table 8 are questions 1 and 3. Hence $j_1 = 1$ and 2 corresponds to $j = 1$ and 3 and $j_2 = 1$ and 2 corresponds to $j = 2$ and 4 in model (8). The expected numbers for this model are shown in column 8 of Table 8. The fit to the data in column 5 is obviously very good.

Because the two questions influenced by latent variable θ_1 are identical, as are the two questions influenced by θ_2, we may want to test the model above with

$$p^{(1)}_{11} = p^{(1)}_{21}, \qquad p^{(1)}_{12} = p^{(1)}_{22}$$
$$p^{(2)}_{11} = p^{(2)}_{21}, \qquad p^{(2)}_{12} = p^{(2)}_{22}$$

We call this Model IIIb. It is illustrated in Figure 4 by connecting the membership questions as well as the attitude questions. The estimates under Model IIIb are shown in Table 9 next to the parameter estimates of Model IIIa. The expected numbers under Model IIIb as shown in Table 8, column 9, indicate a fit much less satisfactory than for Model IIIa. This probably means that the second time the question is asked it is perceived differently.

We shall finally study Model IIIc, which corresponds to (15) with $p^{(1)}_{j_1}(\theta_1)$ and $p^{(2)}_{j_2}(\theta_2)$ given by (17) and (18) and with $\varphi(\theta_1, \theta_2)$ being a two-dimensional normal density. It is in a formal sense identical with Model II, but θ_1 and θ_2 corresponds to the two questions involved instead of time points one and two. This means that the latent variable does not change over time, but instead changes with the question asked.

This is in fact the same model structure as we studied in Models IIIa and IIIb. Since (15) allows for a correlation between θ_1 and θ_2, we can, however, by means of Model IIIc study the extent to which the latent variable influencing the membership questions is correlated with the latent variable influencing the attitude questions. The parameter estimates of Model IIIc are shown below. Note that as for Models IIIa and IIIb questions 2 and 3 have changed place:

$$\hat{\varepsilon}_{11} = -0.189$$
$$\hat{\varepsilon}_{12} = +0.189$$
$$\hat{\varepsilon}_{21} = -0.088$$
$$\hat{\varepsilon}_{22} = +0.088$$
$$\hat{\mu}_1 = +0.995$$
$$\hat{\sigma}_1^2 = +9.372$$
$$\hat{\mu}_2 = -0.302$$
$$\hat{\sigma}_2^2 = +2.189$$

and

$$\hat{\rho} = +0.321$$

The expected numbers based on these estimates are shown in Table 8, column 10. The fit is almost as good as for Model IIIa.

In Table 10 the five models discussed above are compared by means of the goodness-of-fit test statistic (21). It follows from Table 10 that we

TABLE 10
Comparison of Five Models Fitted to the
Coleman Data

Model	q	df	Significance[a] probability
I	761.5	10	<0.001
II	812.6	8	<0.001
IIIa	1.28	4	0.863
IIIb	40.56	8	<0.001
IIIc	5.00	8	0.756

[a] $P(Q \geq q)$, where $Q \sim \chi^2(\text{df})$.

TABLE 11
Obtained Score Vector under Model IIIc and
Assigned Latent Class under Model IIIa

| | Scores | Assigned latent class | | | |
		1	2	3	4
$r_1=0$	$r_2=0$				554
	1		281		338
	2	531			
1	0				184
	1	75	49	85	56
	2	294			
2	0				182
	1	140		171	
	2	458			

have excellent fits for both Models IIIa and IIIc. One criterion that obviously favors IIIc is a search for models with as few parameters as possible. Also an assumption of continuous range spaces for θ_1 and θ_2 seems more realistic than an assumption of only two possible values for θ_1 and θ_2. It is true, on the other hand, that Models IIIa and IIIc are basically identical in structure and should, therefore, describe the data in a similar way. One way to check this is shown in Table 11, where the assigned latent class under Model IIIa is compared with the score vector (r_1, r_2), where r_1 is the score on the membership questions and r_2 is the score on the attitude questions. Under Model IIIc r_1 and r_2 are jointly sufficient statistics for θ_1 and θ_2.

The connection between the score vector and the assigned latent class is not as strong as in Table 7, but we note that the classification indicated by boxes in Table 11 accounts for 81% of the sample.

5. REFERENCES

Andersen, E. B. (1980). *Discrete statistical models with social sciences applications.* Amsterdam: North-Holland.

Andersen, E. B. (1982). Latent trait models and ability parameter estimation. *Applied Psychological Measurement, 6,* 445–461.

Bishop, Y. M. M., Fienberg, S. E., & Holland, P. W. (1975). *Discrete multivariate analysis: Theory and practice.* Cambridge, MA: MIT Press.

Darroch, J. N., Lauritzen, S. L., & Speed, T. P. (1980). Markov fields and log-linear interaction models for contingency tables. *Annals of Statistics, 8,* 552–539.

Goodman, L. A. (1978). *Analyzing Qualitative/Categorical Data: Log-linear Models and Latent Structure Analysis.* London: Addison & Wesley.

Poulsen, C. S. (1981). Latent class analysis and consumer complaining behaviour. In A. Höskuldsson *et al.* (Eds.), *Symposium in applied statistics.* Copenhagen: NEUCC.

PART IV

APPLICATION STUDIES

CHAPTER TEN

Latent Variable Techniques for Measuring Development

JOHN R. BERGAN

1. INTRODUCTION

The seminal work of Lazarsfeld (1950a, b) on latent structure analysis carried out more than three decades ago charted a new direction for research involving relationships between latent and manifest variables. The two branches of investigation stemming from Lazarsfeld's early work, the latent class branch and the latent trait branch, can be used to address a variety of problems of importance in measuring development. This chapter examines latent class and latent trait models in the context of a large-scale application involving the measurement of development in children participating in the Head Start program. The chapter reviews the application of latent class models describing the ordering of skills in a developmental sequence, and it examines the question of determining the effects of one skill on another. Then, the application of latent trait models for testing hypotheses about developmental sequences is discussed. Hierarchical models are reviewed that constrain slope and difficulty item parameters to test hypotheses about difficulty ordering and slope uniformity for item sets reflecting developmental sequences.

JOHN R. BERGAN • Program in Educational Psychology, College of Education, University of Arizona, Tucson, Arizona 85721.

2. MEASUREMENT IN HEAD START:
A PATH-REFERENCED APPROACH

The application of latent variable technology in the Head Start measures program was driven by the special measurement needs of Head Start. The following paragraphs summarize the Head Start measures program, the special requirements that led to the initiation of the program, the approach adopted to measure cognitive functioning in Head Start, and the ways in which latent variable techniques contributed to the measurement effort.

2.1. OVERVIEW OF THE HEAD START MEASURES PROGRAM

In 1981, the University of Arizona received a contract from the Administration for Children, Youth, and Families to develop measures in the cognitive domain for the National Head Start organization. Two subcontractors, the University of Indiana and the University of California at Santa Cruz also participated in the development process. The charge was to construct measures that were developmental in character, that were appropriate for use with the culturally diverse Head Start population, and that could be used in educational planning for the children in the program. During the period between 1981 and 1984, measures were developed in six content areas: language, math, nature and science, perception, reading, and social development. Items were pilot tested and refined in the spring of 1982. The following fall a national field test of the measures was undertaken in 19 Head Start programs. Further refinements were made and additional field testing was undertaken in the spring of 1983. An initial battery containing six scales was prepared in the fall of 1983 and pilot implementation of the entire battery was undertaken in the spring of 1984. In the fall of 1984, the *Head Start Measures Battery* (Bergan & Smith, 1984) was made available for use in Head Start. Currently, the measures are being used with over 21,000 children in 224 Head Start programs around the country.

2.2. RATIONALE FOR A PATH-REFERENCED APPROACH

Early in the project, it became apparent that existing procedures for describing children's performance on tests would not be appropriate to the task of constructing developmental measures that could be used in educational planning. We reasoned that to use developmental assessment information in planning it would be necessary to plan learning experiences based on the child's position in a developmental sequence.

Moreover, we assumed that effective planning would require information about the specific kinds of skills that children occupying various positions in a sequence could be expected to perform. Norm-referenced assessment was ruled out for describing children's performance because it describes ability in terms of position in a norm group rather than position in a developmental sequence. Moreover, it does not link ability to the performance of specific skills, which limits its usefulness in educational planning.

Criterion-referenced assessment initially appeared to be an attractive alternative to the norm-referenced approach. Criterion-referenced assessment is designed to provide information on the kinds of skills that an individual can perform (see, for example, Popham, 1980), which enhances its usefulness in educational planning. Moreover, early writings on criterion-referenced assessment (e.g., Glaser, 1963; Glaser & Nitko, 1971) stressed skill sequencing, which afforded a link between the criterion-referenced approach and the task of indicating the position of an individual in a developmental sequence. Unfortunately, the current criterion-referenced assessment practice of describing an individual's performance in terms of the proportion of items mastered in a well-defined domain (e.g., Hambleton, Swaminathan, Algina, & Coulson, 1978) limited the usefulness of criterion-referenced assessment with respect to the needs of the Head Start measurement program. Describing a child's performance in terms of the proportion of items mastered may restrict the amount of information provided about specific skills. In addition, it offers no information about the sequencing of skills.

A more fundamental problem with the criterion-referenced approach is that it is founded in a psychological perspective that does not afford a rationale for relating the overt performance of skills to the cognitive abilities underlying skill performance. Criterion-referenced assessment makes no use of the ability concept. Criterion-referenced assessment comes out of the behavioral tradition in psychology (see, for example, Hively, Patterson, & Page, 1968). The idea that a latent ability underlies the performance of specific tasks associated with that ability is not included in the behavioral viewpoint. Criterion-referenced assessment affords no theoretical basis for determining that a set of skills form a developmental sequence reflecting an underlying ability. Any arbitrarily chosen set of skills can be organized into a well-defined domain within the criterion-referenced assessment framework (see, for example, Hambleton & Eignor, 1979). For purposes of assessment in Head Start, we wanted an approach that could relate underlying cognitive abilities to overt test performance. We reasoned that an approach explicitly recognizing the role of underlying cognitive functioning in controlling overt

test performance could take advantage of the large and rapidly ac-
cumulating body of research in cognitive development and in so doing
ultimately provide a better understanding of children's thinking than
would otherwise be possible.

2.3. Path-Referenced Assessment and Latent Variable Models

What was needed for the Head Start measures program was a way to
relate a child's ability or developmental level to his or her position in a
developmental sequence or path comprised of an ordered set of skills
reflecting the ability being assessed. A path-referenced approach to
assessment was adopted to meet this need. The term *path-referenced
assessment* refers to the strategy of referencing a child's ability estimated
using latent trait techniques to position on a development path.

Latent class and latent trait techniques played a key role in the
development of the path-referenced approach used in the *Head Start
Measures Battery*. Latent class models were used in the initial phases of
measures construction to validate hypothesized developmental sequences. Latent trait models were initially used to estimate ability
represented by a developmental level score and to relate developmental
level to position in a developmental sequence. Latent trait models were
also used to place items reflecting a broad age span (3–6) on a common
scale. Finally, we have recently begun to explore the possibility of using
latent trait models to test hypotheses about the sequencing of skills.

3. LATENT CLASS MODELS

Latent class models were used in the Head Start measures program
because they provide a way to test hypotheses regarding relationships
between tasks. For example, latent class procedures can be used to test
the hypothesis that two tasks each represented by a set of items are
hierarchically ordered. In the early phases of the Head Start measures
program, models were constructed of the hierarchical structure of tasks in
each of the six content areas for which measures were developed. Pairs of
items were developed representing each task to be investigated. Then
latent class procedures were applied to investigate relationships between
pairs of tasks. The following discussion details the latent class models
used in testing hypotheses about developmental sequences.

3.1. The General Model

All of the latent class models presented in the chapter are variants of
a general unrestricted latent class model, which relates manifest variables

to a latent variable containing a number of latent classes. The general model and all of the variants will be presented for the case of four dichotomous manifest variables. However, all of the models presented can be generalized to more than four variables. The general model expresses cell probabilities in terms of a latent variable composed of T latent classes. In particular, the probability for cell $ijkl$ ($i = 1\text{–}I$, $j = 1\text{–}J$, $k = 1\text{–}K$, $l = 1\text{–}L$) is given by

$$\pi_{ijkl} = \sum \pi^{ABCDX}_{ijklt} \qquad (1)$$

where π_{ijkl} is the probability for cell $ijkl$ and π^{ABCDX}_{ijklt} is the joint probability of pattern $ijkl$ and latent class t. The joint probability may be expressed as

$$\pi^{ABCDX}_{ijklt} = \pi^{X}_{t}\pi^{\bar{A}X}_{it}\pi^{\bar{B}X}_{jt}\pi^{\bar{C}X}_{kt}\pi^{\bar{D}X}_{lt} \qquad (2)$$

where π^{X}_{t} is the probability of latent class t, $\pi^{\bar{A}X}_{it}$ is the conditional probability that variable A will be responded to at level i given latent class t, and the other conditional probabilities are defined similarly. The general model assumes mutual independence among the manifest variables within latent classes. The model also assumes that latent class probabilities must sum to 1 and that conditional response probabilities must sum to 1.

3.2. Model Variations

Model variations may be produced by altering T, the number of latent classes. Various kinds of restrictions can also be imposed on the general model to produce model variations. The models presented in this chapter will make extensive use of restrictions on conditional response probabilities. In particular, sets of conditional response probabilities will be restricted to be equal both within and across latent classes. For example, in a given model the probability of responding correctly to item A in Latent Class 1 might be set equal to the probability of responding correctly to item A in Latent Class 2. The models to be discussed below will also involve restrictions on latent classes. For example, for a particular model the probability of occurrence of Latent Class 2 might be set equal to the probability of occurrence of Latent Class 3.

3.3. Model Fitting

Latent class models relate cell probabilities for response patterns in a contingency table to model parameters including latent class probabilities and conditional response probabilities. Because the model parameters are generally not known, they must be estimated from the data. Goodman (1974a, b) has presented a parameter estimation procedure based on the iterative proportional fitting algorithm that has the attractive feature of avoiding matrix inversion. Clogg (1977) has constructed a computer program that carries out the iterative process necessary to estimate model parameters using the Goodman algorithm. Clogg's program was used to estimate parameters for the models presented in this chapter.

Estimates of cell probabilities may be converted to expected cell frequencies by multiplying them by sample size. The fit of a given model to the data can then be assessed by examining the correspondence between expected and observed frequencies using the chi-squared statistic. Given that the model is identified (see Goodman, 1974b, for a discussion of identification), the degrees of freedom for the chi-squared test can be determined by subtracting the number of parameters to be estimated from the number of response patterns for the items.

It is possible to conduct statistical tests comparing hierarchically related models to arrive at a preferred model. A preferred model is one that improves on the fit offered by other models and provides an acceptable fit for the data. Two models are hierarchically related if one contains all of the parameters of the other plus one or more additional parameters (Bergan, 1983). In some cases hierarchical comparisons are made between two models, neither of which provides an acceptable fit for the data. Comparisons of this kind may provide useful information, but they do not offer a preferred model since a model affording an adequate fit for the data is lacking. The likelihood ratio statistic (L^2) is particularly useful for conducting hierarchical model tests since it can be partitioned exactly (Bishop, Fienberg, & Holland, 1975). The likelihood ratio statistic was used for all of the models examined in this chapter.

3.4. Testing Developmental Hypotheses with Latent Class Models

The models applied to assess development in the Head Start measures program are a subset of a larger group of models that may be used to test developmental hypotheses. Bergan and Stone (1985a) conducted a detailed examination of models of this kind. They discussed models for representing hierarchically ordered skill sequences and related those models to developmental models depicting transition states between nonmastery and mastery classes. They also reviewed models for distinguishing between transition states and measurement error. Finally, they discussed the relationship between models for hierarchical ordering and models representing skills in separate knowledge domains. The interested reader should consult the Bergan and Stone article for information on models not considered in the context of the Head Start application.

In examining the relationship between skills, it may be of interest to determine not only the ordering of skills, but also the effect that one skill has on the performance of another. The Bergan and Stone (1985a) review did not examine the relationship of the models they considered to the problem of determining the effects of one skill on another. However, the models discussed here have implications for determining effects. The following paragraphs present models used in the Head Start measures program and illustrate their application in determining hierarchical ordering and in assessing the effects of one skill on another.

3.4.1. Models for Determining Hierarchical Ordering

The first model to be presented is an unrestricted two-class model that has been used widely in latent class investigations (see, for example, Goodman, 1974a; Macready & Dayton, 1977). This model is used as a bench mark against which to test the assumption that skills are ordered. If a model asserting ordering does not improve significantly on the fit offered by the two-class model, ordering cannot be assumed. The second model (H_2) is a variation of a model proposed by Dayton and Macready (1976) to assess prerequisite ordering. Two skills are assumed to be prerequisitely ordered when there is a class of individuals who have mastered the subordinate skill and are nonmasters of the superordinate skill. In addition, prerequisiteness requires that there be no individuals who fall in the category of being masters of the superordinate skill while being nonmasters of the subordinate skill (Bergan, 1983; Bergan & Stone, 1958a). The third model (H_3) is a four-class model that provides a test for the prerequisiteness assumption and that can be used to depict the situation in which two skills are in separate knowledge domains.

The data used in model testing reflect the performance of Head Start children on items from the math scale of the *Head Start Measures Battery* (Bergan & Smith, 1984). Data set 1 in Table 1 presents these data. The data describe children's performance on four math items. Item A required identifying the numeral 3. Item B called for identifying the numeral 4. Item C required the child to match a group of blocks to the numeral 3, and item D involved matching a group of blocks to the numeral 4.

TABLE 1
Cell Counts for Two Data Sets

Patterns[a]				Counts	
A	B	C	D	Set 1	Set 2
1	1	1	1	197	234
2	1	1	1	19	7
1	2	1	1	15	67
2	2	1	1	38	38
1	1	2	1	25	22
2	1	2	1	15	0
1	2	2	1	15	6
2	2	2	1	30	16
1	1	1	2	30	188
2	1	1	2	4	15
1	2	1	2	13	174
2	2	1	2	34	124
1	1	2	2	43	22
2	1	2	2	13	5
1	2	2	2	30	38
2	2	2	2	71	34

[a] The numbers 1 and 2 indicate passing and failing responses respectively. 1: Item A, identifying the numeral 3; item B, identifying the numeral 4; item C, matching a group of blocks with the numeral 3; item D, matching a group of blocks with the numeral 4. 2: Item A, counting to 10; item B; counting to 10 starting from 6; item C, adding 5 blocks and 2 blocks; item D, word addition (adding 3 plus 1).

The results of model testing for the data set 1 are shown in Table 2. Model H_1 given in the table is the unrestricted two-class model. This model assumes a single latent variable composed of two latent classes, a mastery class and a nonmastery class. The model holds that all four items belong in the same knowledge domain and that development proceeds from nonmastery to mastery of the domain. Model parameters are shown in Table 3. Ten parameters must be estimated under model H_1. There are eight conditional response probabilities (two for each item) associated with each latent class. Four of these are determined since the two conditional response probabilities for each item must sum to 1. Thus, four conditional response probabilities must be estimated for each latent class. One of the two latent class probabilities must be estimated under the model. The other is determined since the latent class probabilities

TABLE 2
Hypothesis Testing Results

Model	L^2	df	P
H_1: Two class unrestricted	66.89	6	<.01
H_2: Three class	18.23	5	<.01
H_3: Four class (related domains)	8.32	4	>.05
H_4: Orthogonal domains (not tested)	—	—	—

must also sum to 1. One parameter must be allotted to reflect sample size. Because the number of response patterns in the table is $2^4 = 16$, there are six degrees of freedom for this model. The L^2 in Table 2 for model H_1 indicates that this model provides an extremely poor fit for the data and thus can safely be rejected.

Model H_2 is a three-class model that assumes that the four items

TABLE 3
Parameter Estimates for Hypotheses[a]

Model/t	Data set 1				
	$\P^X{}_t$	$\P^{\bar{A}X}{}_{1t}$	$\P^{\bar{B}}{}_{1t}$	$\P^{\bar{C}X}{}_{1t}$	$\P^{\bar{D}X}{}_{1t}$
H_1					
1	.441	.954	.991	.878	.862
2	.559	.359	.263	.364	.389
H_2					
1	.344	.931	.971	.999	.980
2	.470	.274	.149	.377	.397
3	.186	.931[b]	.971[b]	.377[b]	.397[b]
H_3					
1	.363	.944	.999	.973	.944
2	.428	.296	.166	.300	.334
3	.139	.944[b]	.999[b]	.300[b]	.334[b]
4	.070	.296[b]	.166[b]	.973[b]	.944[b]

[a] $\P^X{}_t$ represents the probability for latent class t, and $\P^{\bar{A}X}{}_{1t}$ represent the conditional probability for passing variable A given membership in latent class t. The other conditional probabilities are defined similarly. Any of the conditional probabilities for a failing response ($i = 2$) may be obtained by subtracting the conditional probability for a passing response ($i = 1$) from 1.00.

[b] Can be determined from the other entries according to model specification.

under examination are comprised of two prerequisitely ordered skills. The two numeral identification items comprise the subordinate skill and the two matching items comprise the superordinate skill. Model H_2 is hierarchically related to H_1. Model h_2 contains both of the latent classes in H_1 plus an additional latent class ($t = 3$) reflecting the situation in which children have mastered the subordinate skill but are nonmasters of the superordinate skill. Latent class three includes the assumption that the numeral identification skill has been mastered. Accordingly, the conditional response probabilities for items A and B reflecting numeral identification are set equal to the conditional response probabilities for those items in the mastery class ($t = 1$). Latent class three also includes the assumption that matching a numeral to a group of blocks has not been mastered. Thus, the conditional response probabilities for items C and D in latent class three are set equal to the corresponding conditional response probabilities in the nonmastery class ($t = 2$). These restrictions may be written as follows:

$$\P^{\bar{A}X}13 = \P^{\bar{A}X}11, \quad \P^{\bar{B}X}13 = \P^{\bar{B}X}11; \quad \P^{\bar{C}X}23 = \P^{\bar{C}X}22, \quad \P^{\bar{D}X}23 = \P^{\bar{D}X}22$$

Eleven parameters must be estimated under model H_2. These include the ten estimated under H_1 plus one more for the third latent class. Note that the conditional response probabilities within the third latent class do not have to be estimated since these are set equal to conditional response probabilities in the other latent classes. Model H_2 has five degrees of freedom. The L^2 given in Table 2 for this model indicates that the model does not provide an acceptable fit for the data. Because H_2 is hierarchically related to H_1, the L^2 for H_2 can be subtracted from the L^2 for H_1. The degrees of freedom for H_2 can also be subtracted from the degrees of freedom for H_1. The result is an L^2 of 48.66 with 1 degree of freedom, which indicates that H_2 improves significantly ($p < .01$) on the fit afforded by H_1.

Model H_3 is a four-class model that contains the three latent classes in model H_2 plus one additional latent class. This class represents individuals who have mastered the matching skill, but are nonmasters of the numeral identification skill. Model H_3 serves as a test of the prerequisiteness assumption reflected in model H_2. Recall that the assumption that numeral identification is prerequisite to matching numerals to sets of blocks requires that there be no individuals categorized as having mastered the matching skill and at the same time not having mastered the numeral identification skill. Evidence of a class of masters of the matching skill who are also nonmasters of the numeral identification skill contraindicates prerequisiteness.

Model H_3 includes the same restrictions on conditional response probabilities as those present in H_2. In addition, it contains restrictions on the conditional response probabilities in the fourth latent class. The fourth latent class includes individuals who are masters of the matching skill. Accordingly, the conditional response probabilities for the two matching items in the fourth class are restricted to be equal to the corresponding conditional response probabilities in the mastery class ($t = 1$). Individuals in the fourth class are assumed to be nonmasters of the numeral identification skill. Thus, the conditional response probabilities for the fourth latent class for the numeral identification items are set equal to the corresponding conditional response probabilities in the nonmastery class ($t = 2$). The restrictions for H_3 may be expressed as follows:

$$\P^{\bar{A}X}13 = \P^{\bar{A}X}11, \quad \P^{\bar{B}X}13 = \P^{\bar{B}X}11; \quad \P^{\bar{C}X}13 = \P^{\bar{C}X}12, \quad \P^{\bar{D}X}13 = \P^{\bar{D}X}12$$

$$\P^{\bar{A}X}14 = \P^{\bar{A}X}12, \quad \P^{\bar{B}X}14 = \P^{\bar{B}X}12; \quad \P^{\bar{C}X}14 = \P^{\bar{C}X}11, \quad \P^{\bar{D}X}14 = \P^{\bar{D}X}11$$

Twelve parameters must be estimated under H_3. These include the eleven parameters estimated under H_2 plus one additional parameter for the fourth latent class. Model H_3 has four degrees of freedom. The L^2 for this model in Table 2 indicates that it fits the data adequately. Model 3 is hierarchically related to H_2. Subtraction of the L^2 for H_3 from the L^2 for H_2 yields a significant L^2 of 9.91 with 1 degree of freedom ($p > .01$). Because H_3 fits the data and significantly improves on the fit offered by H_2, H_3 is preferred for this data set.

3.4.2. Effects Involving Latent Variables and Item Responses

Bergen and Stone (1985a) demonstrated that model H_3 is equivalent to a model presented by Goodman (1974a) containing two latent variables. The Goodman model has direct implications for measuring the effects of one latent variable on another. Macready (1982) has also discussed the problem of assessing effects in models related to the measurement of development. The approach presented here based on Goodman's procedures is markedly different from the one used by Macready (1982).

Model H_3 was depicted in the preceding paragraphs as being composed of a single latent variable containing four latent classes. Bergan and Stone (1985a) have shown that H_3 also may be thought of as being comprised of two latent variables, Y and Z, each composed of two latent classes. Given this interpretation, H_3 may be conceptualized as reflecting the situation in which two skills, each represented by a latent variable, belong in two separate knowledge domains. The two-variable representation of H_3 may be expressed as follows:

$$\P_{ijkl} = \sum^{YZ}\sum \P^{YZ}rs\,\P^{\bar{A}YZ}irs\,\P^{\bar{B}YZ}jrs\,\P^{\bar{C}YZ}krs\,\P^{\bar{D}YZ}lrs \tag{3}$$

where $\P^{YZ}rs$ represents the probability of latent variable Y at level r ($r = 1, 2$) and latent variable Z at level s ($s = 1, 2$), $\P^{\bar{A}YZ}irs$ is the conditional probability that the response to item A will be at level i given that latent variable Y is at level r and latent variable Z is at level s. The other conditional response probabilities are defined similarly. Model H_3 assumes that variable Y explains performance on items A and B and that variable Z explains performance on items C and D. Latent variable Y is assumed to have no effect on the performance of items C and D. Likewise, variable Z is assumed to have no effect on performance of items A and B. These assumptions are reflected in the following restrictions:

$$\P^{\bar{A}YZ}1r1 = \P^{\bar{A}YZ}1r2, \qquad \P^{\bar{B}YZ}1r1 = \P^{\bar{B}YZ}1r2$$

$$\P^{\bar{C}YZ}11s = \P^{\bar{C}YZ}12s, \qquad \P^{\bar{D}YZ}11s = \P^{\bar{D}YZ}21s$$

These restrictions are equivalent to those presented for the one-variable version of model H_3. To make this clear, the reader may wish to renumber the latent classes so that Latent Class 2 is interchanged with Latent Class 4. Note that under these restrictions the probability of a passing response for items A and B is the same for each of the two levels of latent variable Z. Thus, changes in the level of Z do not affect the probability of passing items A or B. Likewise, the probability of passing C and D is the same for the two levels of latent variable Y. Accordingly, Y has no effect on the probability of correctly responding to C or D. Table 3 shows the latent class probabilities and conditional response probabilities for the two-variable version of model H_3. These values are used in computing effects under the model.

Insofar as H_3 contains two latent variables, it is reasonable to assign each variable to a separate skill or knowledge domain. However, the two knowledge domains are not entirely independent. Goodman (1974a), using techniques developed for log-linear and logit models (Goodman, 1973), has shown that it is possible to assess the relationship between domains in terms of the effects of one latent variable on the other. The association between the two latent variables becomes apparent when one

TABLE 4
Cross Classification of Two Latent Variables

		Variable Z	
		1	2
Variable Y	1	$\P^{YZ}11 = \P^{X}1^a$	$\P^{YZ}12 = \P^{X}3$
	2	$\P^{YZ}21 = \P^{X}4$	$\P^{YZ}22 = \P^{X}2$

[a] The $\P^{X}t$ give the latent classes in the one-variable model that are counterparts to the classes in the two-variable model.

considers the table cross-classifying latent variables Y and Z. This is given in Table 4. Because Y and Z are dichotomous variables, this is a 2×2 table. Independence obtains between two dichotomous variables when the cross-product ratio for the cross classification of the variables equals 1. Thus Y and Z are independent only when $(\P^{YZ}11\P^{YZ}22)/(\P^{YZ}12\P^{YZ}21) = 1$.

The cross-product ratio given in the preceding paragraph is equivalent to the following odds ratio: $(\P^{YZ}11/\P^{YZ}12)/(\P^{YZ}21/\P^{YZ}22)$. The direct effect between the two latent variables is a function of this odds ratio (Goodman, 1974a). In particular, the odds of being at Level 1 as opposed to Level 2 on latent variable Z are equal to the square root of the above odds ratio. The population parameters for the odds ratio are generally not known. However, estimates of those parameters can be used to construct an expected odds ratio. The expected odds ratio is easily interpreted and can be used to communicate information regarding the effects of one latent variable on another. For example, consider the expected odds ratio for the math data. The four latent class probabilities estimated under model H_3 are .363, .428, .139, and .070. The odds ratio for these latent class probabilities is $(.363/.139)/(.070/.428) = 15.97$. Thus, the expected odds of being able to match numerals to blocks correctly are over 16 times higher for the class of children who can identify numerals than they are for the class of children who cannot identify numerals. The expected odds making up the odds ratio for the math data are also informative. The odds of being able to match blocks to numerals for the class of individuals who can identify numerals is 2.67. The odds of being able to match blocks to numerals are over 2:1 for the class of children who have mastered numeral identification. On the other hand the odds of matching blocks to numerals are .1635, far less than 1 for individuals who cannot identify numerals. One might think that the effect demonstrated in the above odds ratio would imply that the probability of passing a matching item given a nonmastery level for matching might be affected by mastery level with respect to numeral identification, but this is not the case. The restrictions on conditional response probabilities included in the model hold the probability of passing a particular matching item to be the same for nonmasters of the matching skill regardless of whether or not mastery is assumed for numeral identification. For example, the probability of a nonmaster of the matching skill passing item C, given mastery of numeral identification, is .30. The probability of a nonmaster of the matching skill passing item C is also .30 for nonmasters of numeral identification.

The interpretation implied by the above discussion is that numeral identification affects matching blocks to numerals. However, in the absence of a time difference in measuring performance on items related to the two latent variables, there is generally no way to determine the direction of the effects between latent variables empirically. If one wished to make the argument that matching blocks to numerals influenced numeral identification, then the association between these two latent variables could be used in support of that argument. In determining the direction of the effect, one typically must rely on theory. In some cases it will be reasonable to assert a unidirectional effect and in others it will be reasonable to assert that effects operate in both directions. Finally, in some instances, it will be appropriate only to recognize an association between latent variables, but not to assume that either variable has an effect on the other.

Model H_3 not only can support the assumption that one latent variable affects another, but also is congruent with the view that a latent variable affects responses to items. Thus, it can be assumed that the latent variable for numeral identification affects responses to items A and B, identifying the numeral 3 and identifying the numeral 4. Likewise, it can be assumed that the latent variable for matching blocks to numerals affects responses to the two matching items, items C and D. These effects are a function of the cross-product ratio for the cross classification of a latent variable with an item (Goodman, 1974a). Conditional response probabilities provide the data for this type of cross classification. For example, consider the cross classification of item A with latent variable Y given in Table 5. Note that latent variable Z is depicted as being at level s in this cross classification. Because Z has no effect on item A, latent variable Z can be at either Level 1 or Level 2 without affecting the conditional response probabilities in the table. Thus, s can be either 1 or 2. The cross-product ratio for the effect of latent variable Y on item A is $(.944 \times .704)/(.296 \times .056) = 40.09$. The odds of passing as opposed to failing item A are 40 times higher for children classified as masters of the

TABLE 5
Cross Classification of Item A with Latent Variable Y

		Variable A	
		1	2
Variable Y	1	$\P^{\bar{A}YZ}_{11s} = .944$	$\P^{\bar{A}YZ}_{21s} = .056$
	2	$\P^{\bar{A}YZ}_{12s} = .296$	$\P^{\bar{A}YZ}_{22s} = .704$

numeral identification skill than they are for children who are classified as nonmasters of the numeral identification skill.

3.4.3. Testing the Effects Assumption

Model H_3 assumes that there is an association between the two latent variables representing, respectively, numeral identification skill and matching skill. In order to test this assumption, it is necessary to construct a model assuming no association between the two latent variables. As indicated earlier, in the special case in which the cross-product ratio for the cross classification of the two latent variables is 1, there is no association between the latent variables. Bergan and Stone (1985a) introduced a model, which will be labeled H_4, constraining the cross-product ratio for the estimated latent classes to be 1. This model is hierarchical to H_3. It contains all of the latent classes in H_3 and all of the restrictions on conditional response probabilities given in H_3. In addition, it includes the cross-product ratio restriction. This restriction reduces the number of latent classes that have to be estimated by 1, eventuating in a gain of 1 degree of freedom over the 4 degrees of freedom available under H_3. Accordingly, H_4 has 5 degrees of freedom. In order to obtain maximum likelihood estimates of model parameters for H_4, the cross-product ratio restriction must be in effect during the course of the iterative process used in determining parameter estimates. At present, Clogg's (1977) computer program does not include provisions for the cross-product ratio restriction. Thus, the fit of H_4 to the data was not tested.

3.4.4. Effects and Prerequisite Ordering

The discussion to this point has focused on effects between skills represented by a two-variable model involving the cross classification of four latent classes. It is also of interest to consider the question of effects between prerequisitely ordered skills. This would be of interest if model H_2 were to provide an adequate fit for the data. As discussed above, model H_2, which is a three-class model, contains the assumptions underlying the concept of prerequisite ordering. Model H_2 can be conceptualized as a two-variable model. In this conceptualization, H_2 is a special case of H_3 in which the latent class for $\P^{YZ}21$ is equal to 0. Under this condition, the cross-product ratio for the cross classification of Y and Z cannot be determined. This is the case because computation of the cross-product ratio would involve dividing by 0.

Although it is not possible to estimate the effects of one latent variable on another under model H_2, it is possible to estimate the effects of each latent variable on the item responses. For example, consider again the effects of latent variable Y on the response to item A. As shown above, this effect is given by the cross classification of latent variable Y and item A. This cross classification can be expressed as

$$\P^{\bar{A}YZ}112\P^{\bar{A}YZ}122$$

$$\P^{\bar{A}YZ}212\P^{\bar{A}YZ}222$$

The values for this cross classification given under H_2 are .931, .274, .069, .726. The cross-product ratio for this cross classification is 35.75, indicating that the odds of passing item A are 35 times higher for children who are masters of numeral identification than they are for children who are not. The effects of Y on item B can, of course, be computed in a similar fashion to that illustrated for item A. It is also possible to compute the effects of latent variable Y on item C or D. However, the restrictions imposed under the model ensure that the cross-product ratio for either of these effects will be 1. For example, the cross-product ratio for the effect of Y on C is $(.377 \times .623)/(.377 \times .623) = 1$. This result is consistent with the fact that under the model latent variable Y is assumed to have no effect on the performance of item C or item D. Likewise, latent variable Z is assumed to have no effect on the performance of item A or item B.

In the above examples, effects of a latent variable on an item response under model H_2 have been computed under the assumption of a two-variable model. Effects may also be computed under the assumption that H_2 contains a single latent variable including three latent classes. As in the two-variable case, the computation of effects requires the cross classification of the latent variable and the item for which effects are computed. By way of illustration, consider the cross classification of latent variable X and item A under model H_2 shown in Table 4. The effect of latent variable X on item A determined from this table is a function of the following two cross-product ratios:

$$(\P^{\bar{A}X}11\P^{\bar{A}X}22)/(\P^{\bar{A}X}12\P^{\bar{A}X}21), \qquad (\P^{\bar{A}X}11\P^{\bar{A}X}23)/(\P^{\bar{A}X}13\P^{\bar{A}X}21)$$

The first of these cross-product ratios will necessarily be 1 under the model. The second will have a meaningful interpretation. For example, the second cross-product ratio is $(.931 \times .726)/(.274 \times .069) = 35.75$. This ratio indicates that the odds of passing item A for those in the mastery class are over 35 times higher than the odds of passing A for those in the nonmastery class. Note that this cross-product ratio is exactly the same as the one for the effect of latent variable Y on item A.

3.5. Summary for Latent Class Models

The application of latent class models illustrated above reflects a hypothesis-testing approach to test development, which stands in sharp contrast to traditional approaches to test construction. As Embretson (1985) has noted, test construction generally does not include hypothesis-testing activities associated with task characteristics. Item writing is more of an art than a science. Those who engage in it are typically given only general guidelines for determining the stimulus characteristics to be included in items. The test developer is satisfied as long as ability can be estimated with an adequate degree of accuracy. Yet attention to task characteristics is crucial in determining the cognitive processes that underly ability (Embretson, 1985). To understand the nature of ability, it is essential to understand the nature of the tasks included in ability tests. This will require testing hypotheses about the effects of task characteristics on test performance.

In the case of the *Head Start Measures Battery*, the central issue to be examined is that of the developmental ordering of tasks. It is assumed that tasks impose demands (Newell & Simon, 1972) on cognitive functioning that affect task difficulty. Hypothesis testing involves the specification of task demands and the effects that they are assumed to have on skill ordering. The latent class models presented above provide one approach to testing hypotheses about skill ordering.

4. LATENT TRAIT MODELS

The *Head Start Measures Battery* is based on a latent ability model of development, which links latent trait constructs to psychological constructs that describe cognitive growth. The constructs of developmental level and developmental path are particularly important in the model. Developmental level indicates degree of growth. It is an individual characteristic that can be represented by the latent ability parameter in a latent trait model. A developmental path is a set of ordered competencies that comprise a developmental sequence. Developmental paths are defined by task characteristics. Paths can be described in terms of the item parameters in a latent trait model.

Latent ability or developmental level may be thought of as a composite of the specific competencies that individuals possess. It is assumed to be composed of cognitive procedures (Anderson, 1980; Brown & Burton, 1978) that govern the performance of specific tasks. A cognitive procedure is a set of actions or processes performed on objects to achieve a goal. The procedure concept links the composite of competencies that make up developmental level to an organized set of processes implemented in the performance of cognitive tasks. The specification of process indicates how competence is achieved, thereby increasing understanding of the nature of competence.

A cognitive procedure generally governs the performance of a large number of tasks. For instance, procedures such as counting and addition are used in performing many tasks. The ideal procedure includes all of the processes necessary to perform tasks falling within the purview of the procedure. Various versions of a procedure invariably exist that are less than ideal. These cannot be observed directly, but must be inferred from observed performance. Latent trait models can be used to infer the particular version of a procedure that an individual currently employs. Knowledge of an individual's developmental level and knowledge of item characteristics make it possible to estimate the probability that an individual will be able to perform tasks reflecting the procedure. For instance, if it is known that a child has a low probability of performing tasks involving regrouping, then it can be assumed that regrouping is not included in the version of the subtraction procedure that the child is currently using.

Cognitive procedures may be described in terms of attributes associated with demands on cognitive functioning. Such description characterizes procedures in terms of processes associated with task difficulty. Variations in task difficulty provide a basis for developmental sequences associated with a given ability.

The latent trait concept of item difficulty plays a key role in defining developmental paths in the model. Ordering is based on the relative difficulty of items reflecting tasks in the progression. Latent trait technology makes it possible to measure item difficulty in terms of the same continuous scale as that used to measure developmental level. Points on the scale indicate path positions.

As indicated in the first section of the chapter, in the early phases of the Head Start measures program, latent class models were used to test hypotheses about the ordering of skills in developmental sequences. Latent trait models were then used to estimate developmental level and to provide a metric for developmental paths. This procedure was not entirely satisfactory in that the technology used to test hypotheses about developmental sequences was not the same as the technology used to relate developmental level to path position. We have recently begun to explore procedures using technology developed by Thissen (1985) to apply latent trait models to test hypotheses about developmental sequences (Bergan & Stone, 1985b). The discussion that follows presents procedures for testing developmental hypotheses using latent trait techniques. Discussion will be limited to the case of dichotomous items.

4.1. THEORETICAL FRAMEWORK

Latent trait models assert that performance on a set of items is explained by one or more latent traits, each of which reflects a continuous scale. The three most commonly used latent trait models are the one-, two-, and three-parameter logistic models designed for use with dichotomous items (Lord, 1980). The three-parameter model introduced by Birnbaum (1968) may be expressed as follows:

$$P(\theta) = \gamma_j + \frac{1 - \gamma_j}{1 + e^{-\alpha_j(\theta - \beta_j)}} \tag{4}$$

where $P(\theta)$ is the probability of responding correctly to an item given a particular ability level θ, and α, β, and γ are item parameters. The α parameter indicates the slope of the relationship between the probability of a correct response and θ. The β parameter specifies the amount of ability needed to respond correctly with a .5 probability. It indicates item difficulty. The γ parameter gives the lower asymptote of the curve indicating the relationship between the probability of a correct response and ability. It is generally assumed to reflect guessing. The two-parameter model may be constructed by setting the γ parameter to 0. Similarly, the one-parameter model, generally referred to as the Rasch model, may be created from the two-parameter model by holding the α parameter constant across all items.

Model variations beyond those associated with the one-, two-, and three-parameter models can be constructred by imposing various restrictions on the item parameters in a manner similar to that used with latent class models. Restrictions on the α and β parameters are particularly important in testing hypotheses regarding developmental phenomena. Restrictions on the β parameters make it possible to test hypotheses about the difficulty ordering of skills. Restrictions on the α parameters can be used to test hypotheses about the slope of the relationship between ability and the probability of responding correctly. The slopes give the extent to which items relate to the same degree to the latent trait.

To illustrate the use of parameter restrictions, consider the case of two items assumed to reflect two skills ordered by difficulty. The hypothesis that one item is easier than the other can be examined by constructing two models. One of these would allow the β values to vary. The second would restrict the β values for the two items to be equal. The model containing the equality constraint would include one less estimated parameter than the model without this constraint since only one difficulty parameter would have to be estimated for the equality model. The other β value would be determined.

Sets of hierarchically related latent trait models can be specified in essentially the same way as that used with latent class models. The two models above are hierarchically related in that the first contains all of the parameters estimated under the second and one additional parameter, the second β value that must be estimated under the model. Accepting the unrestricted model would support the hypothesis of a difficulty sequence between the two skills. Accepting the restricted model would contraindicate the hypothesized sequence.

Models restricting α values may be constructed in the same fashion as that used for β values. Item subsets hypothesized to be related to the same degree to the latent trait may be restricted to have equal slopes. A restricted model is hierarchically related to the unrestricted model. The unrestricted model contains all of the parameters estimated under the restricted model plus the additional slope parameters that must be estimated. The Rasch model is a special case in which all of the item slopes have been restricted to be equal. There are, of course, many other possibilities. Various subsets of items may be assumed to have the same α values without having to assume that all items have the same values.

Restrictions on slopes are of interest in developmental assessment because in the case of equal slopes the ordering of items in terms of the probability of a passing response is preserved throughout the range of abilities. When slopes vary, they may cross, creating the situation in which the probability of passing one item is lower than the probability of passing another in one part of the ability range and higher in another part of the range. Under these circumstances the interpretation of item difficulty becomes complicated. An item with a low β value does not always imply a higher probability of a passing response than an item with a higher β value.

4.2. METHODS OF ESTIMATION

Various procedures have been developed for estimating item parameters in latent trait models (Bock & Aitkin, 1981; Kolakowski & Bock, 1973; Swaminathan, 1983; Wood, Wingersky, & Lord, 1976). Parameters of models to be discussed here were estimated using the marginal maximum likelihood technique introduced by Bock and Aitkin (1981). The Bock and Aitkin procedure addresses a major parameter estimation problem associated with two- and three-parameter models. In these models, it is not possible to eliminate individual ability from the likelihood by conditioning on a sufficient statistic. This creates a situation in which the number of ability parameters to be estimated increases without limit as the number of examinees increases. Bock and Aitkin (1981) deal with this problem by integrating over the parameter distribution and estimating parameters using the marginal distribution.

The multilog computer program developed by Thissen (1985) produces item parameter estimates using the Bock and Aitkin (1981) algorithm. In addition, Thissen's program is capable of implementing restrictions on α and β parameters of the type described in the preceding paragraphs. Thissen's program was used to produce parameter estimates for the models discussed in this chapter.

4.3. MODEL TESTING

As in the case of latent class models, the likelihood-ratio chi-squared statistic may be used to test the fit of a model to the data. The degrees of freedom are computed by subtracting the number of estimated parameters plus one from the number of score patterns. When the number of items under examination is large, there will be many score patterns with zero counts. Bock and Aitkin (1981) suggest that under these conditions patterns with zero counts should be eliminated in determining degrees of freedom. They also suggest pooling cells with expected counts less than 5. They indicate that $2n$ patterns are needed for a conservative test of fit where n is the number of items.

As with latent class models, the use of the likelihood ratio statistic facilitates hierarchical comparisons among models.

4.4. MODELS FOR TESTING DEVELOPMENTAL HYPOTHESES

Eight models are presented below to illustrate model testing related to developmental hypothesis. The performance of 1000 children ranging

TABLE 6
Results for Latent Trait Models

| | α parameters | | | | β parameters | | | | df | L^2 | P |
	α1	α2	α3	α4	β1	β2	β3	β4			
M_1	2.13	4.37	0.56	0.78	−0.92	−0.02	−3.41	0.65	7	17.8	<.01
M_2	2.95	2.95	0.71	0.71	−0.82	−0.01	−2.76	0.70	9	22.2	<.01
M_3	1.36	1.36	1.36	1.36	−1.14	−0.02	−1.73	0.44	10	107.2	<.01
M_4	1.81	1.81	0.46	0.46	−0.47	−0.47	−1.15	−1.15	11	801.5	<.01
M_5	1.29	1.29	1.34	1.34	−1.45	−0.02	−1.45	0.45	10	144.9	<.01
M_6	2.86	2.86	0.71	0.71	−0.82	0.07	−2.77	0.07	10	72.1	<.01
M_7	2.45	2.45	2.00	2.00	−0.84	−0.02	−1.38	1.59	8	88.0	<.01
M_8	2.91	2.91	0.57	4.49	−0.82	−0.01	−3.31	1.46	7	9.5	>.01

from 34 to 74 months of age on four items from the math scale of the *Head Start Measures Battery* was examined. Their responses are given in Data Set 2 of Table 1. Two of the items were counting items and two were addition items. The first counting item required the child to count forward to 10. The second called for counting to 10 starting with the number 6. The first addition item involved adding two sets of blocks, one containing five blocks and the other containing two blocks. The second addition task called for adding three blocks and one block.

The results of model testing are shown in Table 6. Because all of the items called for a constructed response on the part of the child, guessing was assumed not to be a factor in the children's performance. Accordingly, the first model to be tested (M_1) was the familiar two-parameter model. Eight parameters were estimated under this model, four α parameters and four β parameters. Thus, there were $16 - (8 + 1) = 7$ degrees of freedom for this model. The L^2 of 17.8 ($p < .01$) indicates that this model does not provide an acceptable fit for the data.

The second model examined (M_2) imposed constraints on the slope parameters. The slopes for the two counting items were restricted to be equal. Likewise, the slopes for the two addition items were restricted to be equal. Six parameters were estimated for this model, two α parameters and four β parameters. Thus, there were nine degrees of freedom. The L^2 of 22.2 was significant beyond the .01 level. However, the value was quite close to that obtained for the two-parameter model. The difference L^2 for the comparison of M_1 with M_2 was 4.4 with two degrees of freedom ($p > .10$), indicating that M_1 did not improve on the fit of M_2. Because M_2 is more parsimonious than M_1 and because M_1 did not offer a significant improvement in fit, M_2 was preferred over M_1.

The third model examined (M_3) was the widely used Rasch model.

All four of the slope parameters were constrained to be equal under this model. Accordingly, five parameters were estimated, one α parameter and four β parameters. The L^2 obtained for the Rasch model was 107.2 with 10 degrees of freedom, indicating an extremely poor fit for the data. The Rasch model is hierarchically related to M_2. The comparison of the Rasch model with M_2 yielded an L^2 of 85 with 1 degree of freedom, which is significant well beyond the .01 level.

Models 4, 5, and 6 each included restrictions on the β values designed to test hypotheses about the ordering of items according to the values of the β parameters. M_4 constrained the β values for the two counting items to be equal and the β values for the two addition items to be equal and $\alpha_1 = \alpha_2$, $\alpha_3 = \alpha_4$. The L^2 of 801.5 with 11 degrees of freedom indicates a very poor fit. M_4 is hierarchically related to M_2. Clearly M_2 improved dramatically on the fit offered by M_4. M_5 restricted the β value for counting to 10 to be equal to the β value for the two addition items. The L^2 of 144.9 with 10 degrees of freedom also indicates a poor fit. M_5 is also hierarchically related to M_2, and M_2 improves markedly on the fit afforded by M_5. M_6 constrained the β values for the two counting items to be equal to the β value for the second addition item, which required adding three blocks and one block. The L^2 of 72.1 with 10 degrees of freedom revealed another poor fit. Moreover, the comparison of M_6 with M_2 revealed another dramatic improvement in fit favoring M_2.

Model M_2 was preferred over all of the other models tested to this point. However, it does not offer an acceptable fit for the data. Thissen (personal communication) suggested the inclusion of a guessing parameter for the second addition item. Thissen reasoned that even though adding four blocks and one block calls for a constructed response, the range of numbers in young children's experience that might be chosen for an answer was small enough not to rule out the possibility of guessing. Moreover, the presence of the number 3 in the item stem might have predisposed children capable of counting to respond with 4 even though they did not understand the addition task. Model M_7 was designed to include the guessing parameter suggested by Thissen. M_7 included all of the parameters in M_2 and a guessing parameter for the second addition task. Even with the inclusion of the guessing parameter, an acceptable fit was not obtained ($L^2 = 88.0$, df $= 8$, $p < .01$). The slope restrictions present in M_2 did not yield a fit when the γ parameter was added to the $3 + 1$ addition task. Given the lack of fit for M_7, it was decided to introduce model M_8, which restricted the slopes for the counting items to be equal, but allowed the slopes for the addition items to vary. Model M_8 also included the γ parameter restriction suggested by Thissen. The L^2

for M_8 with seven degrees of freedom was 9.5, indicating an acceptable fit to the data ($p > .10$). Model M_8 is hierarchically related to M_7. The comparison of M_7 with M_8 yielded a difference L^2 of 78.5, suggesting a significant improvement in fit ($p < .01$). Because M_8 fits the data and because it improves significantly on the fit provided by M_7, it was preferred for this data set.

Model M_8 assumes uniform slopes for the counting items. In addition, it asserts that the items are ordered in terms of their β values. The first addition task is the item with the lowest β value. This item is followed in order by the two counting tasks. The second $3 + 1$ addition task with the guessing parameter had the highest β value. The ordering for the two counting tasks is to be expected. The first counting task starts the number chain from 1. The second counting task requires counting on from a number greater than 1. Studies have shown that counting on is more difficult than counting from 1 (e.g., Bergan, Stone, & Feld, 1984; Fuson, 1982). Counting on requires recognition of the cardinal value of the initial number if the final number is to be understood as the number of things that have been counted. This is more difficult than treating the initial number as merely the first number spoken in the chain. The high β value of the verbal addition task is also not surprising. Verbal addition has been found to be superordinate to counting on (Bergan, Stone, & Feld, 1984). Young children often use the counting on strategy as a means for adding. For example, a child may add 5 plus 2 by saying "five" and then counting on "six, seven." The most interesting finding was that the block addition task had the lowest β value of the four items. Addition is generally more difficult than counting. However, addition of two groups of blocks may be carried out by pushing the groups together and counting the resultant pile. Thus, addition with blocks can become a counting task. This task differed from the first counting task in that the child had blocks to count, whereas the first counting task required counting without objects. Previous research has shown that counting with objects is easier than counting without objects (Bergan, Stone, & Feld, 1984). The child who counts blocks is given a concrete representation of the things counted. In counting without objects, the child must construct his/her own concrete representation (e.g., by counting on the fingers) or must represent the objects symbolically.

4.5. Summary and Conclusions for Latent Trait Models

The models presented in the discussion of latent trait techniques demonstrate the use of hierarchical sets of models for testing hypotheses of interest regarding developmental phenomena. The most straightforward hypothesis about a set of items is that they are ordered by difficulty, and that their slopes and lower asymptotes are the same. When it provides an adequate fit for the data, the Rasch model, which constrains slopes to be equal and lower asymptotes to be zero, represents a good starting point for testing hypotheses asserting ordering by difficulty. With the Rasch model as a starting point, the difficulty hypothesis can be tested by examining models restricting β values to be equal for selected subsets of items and then comparing the fit of these models to the fit of one or more models that do not include equality restrictions. The determination of a preferred model in the hierarchical model testing process can be expected to indicate a set of items that may be assumed to be ordered in terms of their β values. Given this state of affairs, the probability of passing an item with a low β value will be less than the probability of passing an item with a high β value throughout the ability range.

The interpretation of findings becomes complex when uniform slopes and/or equal lower asymptotes cannot be assumed across items. When slopes are not uniform, the probability of passing an item with a low β value may not be less than the probability of passing an item with a higher β value throughout the ability range. This state of affairs complicates the interpretation of item difficulty. One item may be harder than another at low ability levels and easier than the other at high ability levels. The complexity of interpretation is lessened somewhat when subsets of items have the same slope, as was the case for the four items examined here. Under these conditions, the interpretation of item difficulty is straightforward within item subsets having uniform slopes. This is the case because the ordering of items within subsets is constant throughout the ability range. Complexity arises only when one wishes to compare items across subsets.

Complexity in interpretation is increased when lower asymptotes are not the same across items. Under these conditions, it is conceivable for two items to have the same β value and yet indicate a different response probability at a given ability level. For example, consider the case in which two items have identical β values and one has a lower asymptote of .23 while the other has a lower asymptote of 0. At $\Theta = 0$, the probability of passing the former item would be .61 while the probability of passing the latter would be .5. Under conditions in which slopes and/or asymptotes are not uniform, interpretation of developmental ordering may be facilitated by indicating the relationship between the probability of a correct response and ability across the ability range.

5. CONCLUSION

The applications of latent variable procedures illustrated in this chapter provide information not only about individuals responding to cognitive tasks, but also about the characteristics of the tasks, and about the relationship between individual characteristics, task characteristics, and task performance. For instance, in the latent trait application, the developmental level score provided information about individual ability. The values for the α and β item parameters offered information about the characteristics of the tasks being performed, and the equation relating ability and item parameters to response probability indicated the effect of individual characteristics and task characteristics on the probability of correct performance.

Applications relating individual characteristics and task characteristics to performance are markedly different from time-honored approaches to educational testing. The fundamental strategy in testing since the time of Galton has been to reference estimates of individual ability to position in a norm group. In the norm-referenced approach, no attempt is made to relate the construct of ability directly to the performance of specific tasks. Tests have generally not provided information about the kinds of tasks individuals of varying ability levels can be expected to perform. Cognitive theorists (e.g., Newell & Simon, 1972) have argued effectively that to understand the nature of an ability is to understand the nature of the tasks that able individuals can and cannot perform. The failure of traditional psychometrics to relate the construct of ability to task performance has cut off a vital source of information for increasing the scientific knowledge base regarding the nature of abilities.

The application of latent variable techniques in testing affords a way to relate abilities, task characteristics, and task performance. As a consequence, latent variable technology represents an important avenue for reestablishing the once strong relationship between psychometrics and cognitive theory and research. A number of investigators have recognized this fact and have begun to propose ways in which latent variable techniques can be applied not only to assess ability, but also to advance the state of knowledge about the construct of ability. Work with latent trait models has been particularly impressive in this regard (see, for example, Embretson, 1985).

The application of latent variable technology in ways that relate latent ability, task characteristics, and task performance has profound implications for education. When test results indicate the kinds of tasks that individuals of varying ability levels can be expected to perform, learning activities can be designed that assist learners to acquire those skills in a developmental sequence that they are ready to learn. The teacher can plan instruction so that the tasks presented to learners are neither too far above the skills associated with their current ability levels nor too far below current levels of functioning. Instruction that eventuates in the acquisition of developmentally related skills can be expected not only to assist in the learning of new skills, but also to aid the learner to advance to a higher level of ability. The goals of instruction, then, need not be restricted to the teaching of isolated skills. Rather they can be targeted directly at the teaching of ability. Under these conditions, specific skills may be organized within the context of the abilities to which they are related. Objectives linked to the teaching of specific skills can be related to long-range educational goals targeted at the development of abilities.

The widespread application of latent variable techniques relating ability, task characteristics, and task performance in education will require an extensive amount of research. It will be necessary to identify the developmental structure of knowledge in the various areas that make school curricula. Information on the ordering of skills can then be used to establish developmental sequences to guide instruction.

Factual knowledge will have to be separated from skills that are developmentally sequenced. The difficulty of items assessing factual knowledge can be expected to vary as a function of instruction. For example, identifying the chemical symbol for table salt may be difficult for those who have not been exposed to instruction, but quite easy for those who have. Differential instruction involving items of factual knowledge can affect the ordering of item difficulties in groups receiving instruction emphasizing different instructional content. When this occurs, the unidimensionality assumption, which is fundamental to latent variable techniques, is violated (Traub & Wolfe, 1981). Test involving hypothesized developmental skill sequences must be monitored continuously to identify instances in which variations in difficulty ordering occur across groups.

6. REFERENCES

Anderson, J. R. (1980). *Cognitive psychology and its implications*. San Francisco: W. H. Freeman.

Bergan, J. R. (1983). Latent-class models in educational research. *Review of Research in Education, 10*, 305–360.

Bergan, J. R., & Smith, A. N. (1984). *Head Start Measures Battery*. Tucson, AZ: University of Arizona, Center for Educational Evaluation and Measurement.

Bergan, J. R., & Stone, C. A. (1985a). Latent class models for knowledge domains. *Psychological Bulletin, 98,* 166–184.

Bergan, J. R., & Stone, C. A. (1985b). *Restricted item response models for developmental assessment.* Paper presented at the annual meeting of the American Educational Research Association, Los Angeles, CA.

Bergan, J. R., Stone, C. A., & Feld, J. K. (1984). Rule replacement in the development of basic number skills. *Journal of Educational Psychology, 76,* 289–299.

Birnbaum, A. (1968). Some latent trait models and their use in inferring an examinee's ability. In F. M. Lord & M. R. Novick (Eds.), *Statistical theories of mental test scores* (pp. 397–424). Reading, MA: Addison-Wesley.

Bishop, Y. M., Fienberg, S. E., & Holland, P. W. (1975). *Discrete multivariate analysis.* Cambridge, MA: MIT Press.

Bock, R. D., & Aitkin, M. (1981). Marginal maximum likelihood estimation of item parameters: Application of an algorithm. *Psychometrika, 46,* 443–459.

Brown, J. S., & Burton, R. R. (1978). Diagnostic models for procedural bugs in basic mathematical skills. *Cognitive Science, 2,* 155–192.

Clogg, C. C. (1977). *Unrestricted and restricted maximum likelihood latent structure analysis: A manual for users.* Unpublished paper, Pennsylvania State University.

Dayton, C. M., & Macready, G. B. (1976). A probabilistic model for validation of behavior hierarchies. *Psychometrika, 41,* 189–204.

Embretson, S. E. (1985). *Test design developments in psychology in psychometrics.* Orlando, FL: Academic Press.

Fuson, K. C. (1982). An analysis of the counting-on solution procedure in addition. In T. P. Carpenter, J. M. Moser & T. A. Romberg (Eds.), *Addition and subtraction: A cognitive perspective.* Hillsdale, NJ: Lawrence Erlbaum Associates.

Glaser, R. (1963). Instructional technology and the measurement of learning outcomes: Some questions. *American Psychologist, 18,* 519–521.

Glaser, R., & Nitko, A. J. (1971). Measurement in learning and instruction. In R. L. Thorndike (Ed.), *Educational measurement* (2nd ed.). Washington, DC: American Council on Education.

Goodman, L. A. (1973). Causal analysis of data from panel studies and other kinds of surveys. *American Journal of Sociology, 78,* 1135–1191.

Goodman, L. A. (1974a). The analysis of systems of quantitative variables when some of the variables are unobservable: Part I. A modified latent structure approach. *American Journal of Sociology. 79,* 1,179–1,259.

Goodman, L. A. (1974b). Exploratory latent structure analysis using both identifiable and unidentifiable models. *Biometrika, 61,* 215–231.

Hambleton, R. K., & Eignor, D. R. (1979). *A practitioner's guide to criterion-referenced test development, validation, and test score usage.* Unpublished manuscript prepared for the National Institute of Education and Department of Health, Education, and Welfare.

Hambleton, R. K., Swaminathan, H., Algina, J., & Coulson, D. G. (1978). Criterion-referenced testing and measurement: A review of technical issues and developments. *Review of Educational Research, 48,* 1–48.

Hively, W., Patterson, H. L., & Page, S. H. (1968). A "universe-defined" system of arithmetic achievement tests. *Journal of Educational Measurement, 5,* 275–290.

Kolakowski, D., & Bock, R. D. (1973). LOGOG: *Maximum likelihood item analysis and test scoring—Logistic model for multiple item responses.* Chicago: National Educational Resources.

Lazarsfeld, P. F. (1950a). The interpretation and computation of some latent structures. In

S. A. Stouffer *et al.* (Eds.), *Measurement and prediction*. Princeton, NJ: Princeton University Press.

Lazarsfeld, P. F. (1950b). The logical and mathematical foundation of latent structure analysis. In S. A. Stouffer *et al.* (Eds.), *Measurement and prediction*. Princeton, NJ: Princeton University Press.

Lord, F. M. (1980). *Applications of item response theory to practical testing problems*. Hillsdale, NJ: Lawrence Erlbaum Associates.

Macready, G. B. (1982). The use of latent class models for assessing prerequisite relations and transference among traits. *Psychometrika, 47,* 477–488.

Macready, G. B., & Dayton, C. M. (1977). The use of probabilistic models in the assessment of mastery. *Journal of Educational Statistics, 2,* 99–120.

Newell, A.. & Simon, H. A. (1972). *Human problem solving*. Englewood Cliffs, NJ: Prentice-Hall.

Popham, W. P. (1980). Content domain specification/item generation. In R. A. Berk (Ed.), *Criterion-referenced measurement: The state of the art*. Baltimore: Johns Hopkins University Press.

Swaminathan, H. (1983). Parameter estimation in item response models. In R. K. Hambleton (Ed.), *Applications of item response theory* (pp. 24–44). Vancouver, BC, Canada: Educational Research Institute of British Columbia.

Thissen, D. (1985). MULTILOG, Version 4.0 User's Guide. Lawrence, KS: University of Kansas.

Traub, R. E., & Wolfe, R. G. (1981). Latent trait theories and the assessment of educational achievement. *Review of Research in Education, 9,* 377–435.

Wood, R. L., Wingersky, M. S., & Lord, F. M. (1976). LOGIST: *A computer program for estimating examinee ability and item characteristic curve parameters* (Research Memorandum 76-6). Princeton, NJ: Educational Testing Service.

CHAPTER ELEVEN

Item Bias and Test Multidimensionality

FRANK KOK

1. INTRODUCTION

In recent decades the question whether tests are equally "fair" for members of several (cultural) groups has received considerable attention. Often only the item responses of one single test measuring the construct of interest are available, which means the researcher does not have a criterion at his disposal by which the "fairness" of the total scores on the test for testees of different groups can be judged. This is one of the reasons why in the last decade the special question whether the individual test items are "biased" has received much attention.

An item is defined as biased if members of different (cultural) groups of the same ability level do not have the same probability of answering the items correctly. Ability refers to the trait most items of the test supposedly measure. Take, for instance, a test that supposedly measures the ability to reason logically and where some items are verbal arithmetic problems. To answer these items correctly, special knowledge of the English vocabulary is required. Members of a minority group may answer these items correctly less frequently than majority group members, even if they have, on the average, an equal level of logical reasoning ability.

Item bias research serves two related purposes. *The first goal* is the identification of items that are to be suspected of bias, in the sense of the definition presented above. The procedure for identifying biased items

FRANK KOK • Psychological Laboratory, University of Amsterdam, 1018 XA Amsterdam, The Netherlands.

consists of the specification of a formal definition for item bias and the construction of a statistical procedure providing an adequate decision rule: item i is probably biased (according to the formal definition) if the test statistic is greater than some critical value.

There seems to be general consensus about the proper formal definition: it is assumed that a latent trait parameter ξ can be assigned to every subject, representing the ability that the test supposedly measures. The population proportion of testees with ability ξ who answer the item correctly can be plotted against ξ, resulting in the so-called item characteristic curve (ICC). An item is unbiased if and only if the ICCs evaluated in the two groups considered coincide.

In recent years many (statistical) techniques have been suggested for the identification of biased items (if item responses of one test only are available). (For reviews see Berk, 1982; and Rudner, Getson, & Knight, 1980.) The main purpose of the detection techniques as they are applied in practice is to decide whether there are substantial differences between the ICCs evaluated separately in the two groups. This dichotomous decision results from applying some bias detection procedures to the "raw data," that is, to the item responses (correct, incorrect, agree, disagree). Many of the detection techniques are not based on the latent trait theory, but test certain alleged consequences of noncoinciding ICCs. For instance, items that are biased in this sense will often appear as outliers, if transformed group items' p values are plotted against each other (see, e.g., Angoff, 1982).

The discovery of noncoinciding ICCs does not necessarily mean that the item is "unfair." For instance, if a test intends to measure verbal ability and some items require knowledge of special concepts that are not taught in black school districts, it remains a point of discussion whether the item is unfair or whether blacks really do have a lower level of verbal ability because of their disadvantaged position. In any case, a judgement about the possible unfairness of an item requires knowledge of the mechanisms underlying the occurrence of noncoinciding ICCs. *The second and most important goal* of item bias research is the explanation of the occurrence of item bias in terms of meaningful, psychological constructs.

A very natural way of "explaining" item bias in nonscientific terms is to state that the test measures for one group something other than for the other group. Hunter (1975) describes item bias in terms of group differences in distributions of secondary latent traits (in the example presented above, verbal ability would be a secondary ability and logical reasoning a primary ability). Linn, Levine, Hastings, and Wardrop (1981) conceptualize bias as multidimensionality, and Green (1982) asserts that future research in item bias will have to concentrate on the unidimensionality issue. Mellenbergh (1984) specifically defines the enterprise of explaining "statistical" bias in terms of psychological constructs as the problem of "finding the biasing traits." The conclusion is that in verbal accounts, item bias is often explained in terms of test multidimensionality.

The statistical models underlying most bias detection procedures do not account for the mechanisms responsible for the occurrence of item bias, because the main objective is the discovery of noncoinciding ICCs. Admittedly some of the proposed statistical bias detection procedures not only pretend to provide information about noncoinciding ICCs, but also about the nature of the differences between the ICCs. Several authors (Marasquilo & Slaughter, 1981; van der Flier, Mellenbergh, Adèr, & Wijn, 1984) describe techniques for testing the hypotheses about differences of success probabilities at different levels of the latent trait continuum. Lord (1980) describes methods for investigating whether differences in location of slopes are responsible for noncoincoding ICCs. A few detection methods have been reported that account for causal factors. Examples are the work of Muthén and Lehman (1985), who describe item bias in terms of different "error" variance, and the work of Veale and Foreman (1983), who present a model for explaining item bias from different patterns of foil responses. The scope of these approaches, however, is only limited regarding the problem of formulating a general model accounting for different mechanisms underlying the occurrence of item bias. One reason for this may be that in the current methodology detecting biased items is a different problem from explaining bias. In the proposals for bias item detection procedures, like those discussed above, usually a mathematical model for item bias is proposed. The application of these procedures requires estimation of parameters.

The behavior of detection procedures is often investigated by computer simulation research. In these studies "biased" tests are generated by the computer, using some mathematical model. In this case the researcher has more freedom, because he does not have to worry about estimation problems. In studies of this kind "biased" tests are often constructed by setting item difficulty or discrimination parameters at different values for two "groups." This procedure does not account for causal mechanisms either. In this chapter a mathematical model for generating item responses in simulation research is presented, that accounts for the bias phenomenon in terms of multidimensionality. This model is to be used in simulation research, so no attention is paid to the problem of estimating parameters. The model seems to be interesting in its own right, because it makes the relation between test multidimensionality and item bias (in the sense of different ICCs) explicit in a quantitative way. It illuminates some special features of the item bias phenomenon as well.

In the next section first the issue of test multidimensionality is discussed, and in the last sections a discussion of some features of the model and its applications is presented.

2. TEST MULTIDIMENSIONALITY AND RELEVANT ABILITIES

In this section it will be argued that different types of ability may be relevant for providing the correct answer to a test question. A distinction will be made between the statement that several abilities are relevant and the statement that the test is multidimensional in a psychometric sense.

The starting point is the notion that the item response success probability depends on the ability the test intends to measure. In the model presented in the next section this ability is represented by the latent trait ξ. In some cases, lack of mastery of this trait can be compensated by other abilities. For instance, when a testee takes a test of Spanish vocabulary, his/her lack of knowledge of the Spanish language can be compensated by mastery of the French language. This type of ability will be referred to as "compensatory" ability and represented by the latent trait η_1. Finally, the probability of a testee answering the item correctly depends on certain abilities that allow him/her to understand the question. A simple example, of course, is knowledge of the language the item is formulated in. In the following, this type of ability will be called conditional ability. It is assumed that a latent trait η_2 can be associated with this ability. If the testee does not know the answer to the question, the probability of answering the item correctly depends on his ability to make use of certain contextual cues the particular item may offer. This fourth type of ability, η_3, will be called *testwiseness*.

According to Lord and Novick (1986, 538) a test is unidimensional, or, in other words, "an individual's performance depends on a single underlying trait if, given his value on this trait, nothing further can be learned from him that can contribute to the explanation of his performance." In latent trait theory this view is expressed by the concept of local stochastic independence: On condition of ability ξ, the items are statistically independent. A test is considered n dimensional in a psychometric sense if stochastic independence between the items is observed only after conditioning on n latent traits. It is important to note that the judgment about the dimensionality of the test is only meaningful in relation to a certain population.

An ability can be considered *relevant* in relation to answering a test item if it covaries with the item response success probability in the *total* population, consisting of all individuals that will possibly take the test (given the level of other relevant latent traits). A test is unidimensional in a certain subpopulation if:

1. No other abilities than ξ are relevant.
2. Some other abilities are relevant but in the subpopulation considered (for instance the white majority group) these abilities do not covary with the item response probabilities. For instance, all white testees have a sufficient knowledge of the native language to understand the question perfectly. In other words, latent trait η_2 as defined previously does not covary in the white population with the item response success probability (given the level of other latent traits).
3. Some abilities other than ξ are relevant, but covary with the response success probability of one item only. In this case, given the testee's value on latent trait ξ nothing further can be learned from him/her that can contribute to the explanation of his performance, because on condition of ξ all items are still statistically independent. The additional ability is comparable to what is called a unique factor in factor analysis.

In general, if n abilities are relevant, the test administered in a specific group can still be k dimensional with $k < n$.

3. MODEL

Latent traits ξ, η_1, η_2, η_3 as defined in the previous section are assumed to represent all relevant abilities. We introduce the (increasing) functions $\phi_{1i}(\xi, \eta_1)$, $\phi_{2i}(\eta_2)$, and $\phi_{3i}(\eta_3)$ for describing the relationship between the separate latent traits and the response success probability of item i.

It is assumed that $\zeta = \xi + \alpha_{2i}\eta_1$ represents the combined influence of ability of interest ξ and compensatory dimension η_1 (α_{2i} indicates the influence of trait η_1 relative to ξ). ϕ_{1i} [range $(0, 1)$] accounts for the influence of ζ. Testees with high values of ζ will have a probability close to unity of answering the item correctly as opposed to testees with low values of ζ, who will have a probability close to zero (if they understand the question and there is no guessing).

ϕ_{2i} [range $(0, 1)$] accounts for the influence of the conditional ability η_2. Testees with very little understanding of the native language, will almost never understand the question as opposed to testees with a very high knowledge of the (native) language.

ϕ_{3i} [range $(0, 1)$] accounts for the influence of testwiseness (η_3). Very testwise testees will, if they do not know the correct answer, provide a positive response with high probability; testees with low values of η_3 will answer the question correctly with low probability.

The probability of answering an item correctly, if the correct answer is unknown, will be the same for equally testwise individuals. Group membership is irrelevant. The probability of understanding the question will be the same for individuals with equal knowledge of the language the item is written in, and will be independent of group membership. Given the level of testwiseness and the level of the conditional ability, the item response success probability will be the same for all testees with the same combined level of the ability the test intends to measure and a compensatory dimension. If $\zeta = \xi$, that is, $\alpha_{2i} = 0$, ϕ_{2i} is the "pure," unbiased unidimensional item characteristic function (of ξ).

In our model the item response success probability for item i in group j is written as a function of all relevant abilities ξ, η_1, η_2, η_3 as follows:

$$P_j(X_i = 1 \mid \xi, \eta_1, \eta_2, \eta_3) = \phi_{3i}(\eta_3) + [1 - \phi_{3i}(\eta_3)]\phi_{2i}(\eta_2)\phi_{1i}(\xi + \alpha_{2i}\eta_1)$$

$$(1)$$

For testees with latent traits η_2, η_3, $\phi_{3i}(\eta_3)$ determines the lower asymptote and $\phi_{2i}(\eta_2)$ determines the upper asymptote of the function $P_j(X_i = 1 \mid \xi, \eta_1)$. Of testees with latent traits ξ, η_1, η_2 a fraction ϕ_{2i} understands the question, and of this proportion a fraction ϕ_{1i} knows the answer.

From the assumption stated previously regarding the functions ϕ_{1i}, ϕ_{2i}, and ϕ_{3i}, it follows that ϕ_{1i}, ϕ_{2i}, and ϕ_{3i} are independent of group membership. Consequently,

$$P_1(X_i = 1 \mid \xi, \eta_1, \eta_2, \eta_3) = P_2(X_i = 1 \mid \xi, \eta_1, \eta_2, \eta_3) \overset{\text{def}}{=} \phi_i(\xi, \eta_1, \eta_2, \eta_3)$$

$$(2)$$

As stated previously, an item is considered to be unbiased if testees of the same ability level randomly sampled from different groups have the same probability of answering the item correctly:

$$P_1(X_i = 1 \mid \xi) = P_2(X_i = 1 \mid \xi)$$

where ξ is the ability of interest. From (2) and from basic statistical theorems it follows that

$$P_j(X_i = 1 \mid \xi) = \iiint P_j(X_i = 1 \mid \xi, \eta_1, \eta_2, \eta_3)$$

$$\times \, h_j(\eta_1, \eta_2, \eta_3 \mid \xi) \, d\eta_1 \, d\eta_2 \, d\eta_3 \qquad (3)$$

$$= \iiint \phi_i(\xi, \eta_1, \eta_2, \eta_3) h_j(\eta_1, \eta_2, \eta_3 \mid \xi) \, d\eta_1 \, d\eta_2 \, d\eta_3$$

4. DISCUSSION OF THE MODEL

The basic assumption underlying item bias model (3) is that a test measures the same for all groups. Item bias can possibly occur when the test results are erroneously interpreted to represent a unidimensional ability. Because ϕ_i in (3) is the same function for both groups, a necessary condition for the occurrence of item bias is

$$h_1(\eta_1, \eta_2, \eta_3 \mid \xi) \neq h_2(\eta_1, \eta_2, \eta_3 \mid \xi) \qquad (4)$$

Item bias is conceived as a possible consequence of unequal conditional distributions of additional abilities in both groups. This model conforms very well with possible psychological explanations of item bias.

Moreover, formula (1) makes clear that additional abilities may influence the response success probability in a number of different ways. Consequently, group differences of item characteristic curves of a different nature may occur depending on what kind of additional ability influences item responses. It should be noted that a test that is multidimensional in a psychometric sense is not necessarily biased. Even if the test is four dimensional

$$P_1(X_i = 1 \mid \xi) = P_2(X_i = 1 \mid \xi)$$

if

$$h_1(\eta_1, \eta_2, \eta_3 \mid \xi) = h_2(\eta_1, \eta_2, \eta_3 \mid \xi)$$

Because of (3) and (2), Item i is unbiased. Moreover, a test that is unidimensional in a psychometric sense is possibly biased; if (4) applies and η_1, η_2, η_3 are latent traits causing unique response variance, then P_j from (3) will generally not be equal for $j = 1, 2$, while the test appears to be unidimensional both in either group and in the total group.

For purposes of illustration a special case of the model will be considered. Firstly, assume that in (3) ξ, η_1, η_2, and η_3 are statistically independent. (This assumption is admittedly very unrealistic.) Now from (3) and (1) it follows that the probability of answering an item correctly in group j can be written as

$$P_j(X_i = 1 \mid \xi) = \int \phi_{3i}(\eta_3)h_j(\eta_3)\,d\eta_3 + \left[1 - \int \phi_{3i}(\eta_3)h_j(\eta_3)\,d\eta_3\right]$$

$$\times \int \phi_{2i}(\eta_2)h_j(\eta_2)\,d\eta_2 \times \int \phi_{1i}(\xi + \alpha_{2i}\eta_1)h_j(\eta_1)\,d\eta_1 \quad (5)$$

$$= R_{ji} + (1 - R_{ji}) \times C_{ji}\int \phi_{1i}(\xi + \alpha_{2i}\eta_1)h_j(\eta_1)\,d\eta_1$$

For group j R_{ji} and C_{ji} appear as item parameters. These can be different for Group 1 and Group 2. Possible interpretations of different values of C_{ji} and R_{ji} are illustrated by some examples.

In general the sizes of the constants R_{ji} and C_{ji} depend on the distribution of η_2 and η_3 in group j and on the shape of the functions ϕ_{3i} and ϕ_{2i}. Suppose η_2 represents the ability to understand the native language and $\phi_2(\eta_2)$ represents the probability to understand the meaning of the question, given the level of knowledge of the native language. Suppose further that within population j all testees understand the native language to a sufficient degree to understand the meaning of the question perfectly well. Formally this can be written as

$$\int_{\eta_{20}}^{\infty} h_j(\eta_2)\,d\eta_2 = 1 \quad \text{and} \quad \phi_{2i}(\eta_2) = 1 \qquad \text{for } \eta_2 > \eta_{20}$$

Now it follows that

$$C_{ji} = \int_{\eta_{20}}^{\infty} [1h_j(\eta_2)]\,d\eta_2 = 1$$

Of course, it is possible that in population j certain testees do not master the native language to a sufficient degree to understand the meaning of the question. In this case, $\phi_{2i} < 1$ for some η_2 occurring in population j and

$$C_{ji} = \int_{-\infty}^{\infty} \phi_{2i}(\eta_2)h_j(\eta_2)\,d\eta_2 < 1$$

Suppose that testees with levels of testwiseness between η_{30} and η_{31} choose at random between the k alternatives of multiple choice item i (if they do not know the answer) and testees with levels smaller than η_{30} are particularly drawn to some attractive, wrong alternative, whereas testees with levels greater than η_{31} can correctly figure out which alternatives can be eliminated. Now, if in the population the density $h_j(\eta_3) = 0$ for $\eta_3 < \eta_{30}$ and $\eta_3 > \eta_{31}$, then $\phi_{3i}(\eta_3)$ will be $1/k$ for the range of η_3 occurring in the population. It follows that

$$R_{ji} = \int_{\eta_{30}}^{\eta_{31}} (1/k)h_j(\eta_3)\,d\eta_3 = 1/k$$

If some people have levels of testwiseness greater than η_{31} or, alternatively, smaller than η_{30} then $\phi_{3i}(\eta_3)$ will be greater and smaller than $1/k$ in the respective population ranges. The values of R_{ji} will be greater or smaller than $1/k$ dependent on the distribution of η_3 in population j.

If $C_{ji} = 1$ in population j, $\alpha_{2i} = 0$ and ϕ_{1i} is the two-parameter logistic function then

$$P_j(X = 1 \mid \xi) = R_{ji} + (1 - R_{ji})\{1 + \exp[-\alpha_{1i}(\xi - \sigma_i)]\}^{-1} \int h_j(\eta_1)\,d\eta_1 \tag{6}$$

$$= R_{ji} + (1 - R_{ji})\{1 + \exp[-\alpha_{1i}(\xi - \sigma_i)]\}$$

which is the three parameter logistic function.

In (5) and (6) α_{1i} and α_{2i} are introduced as pure item parameters independent of group membership. Sometimes it is observed that in two groups items discriminate differently with respect to the ability of interest. This phenomenon can be accounted for by different distributions of η_1. In Figure 1 a special case of (5) is presented where ϕ_{1i} is the two-parameter logistic function with parameters σ_i and α_{1i} and $\alpha_{1i} = 0.46$, $\sigma_i = 0$, $\alpha_{2i} = 1$, $R_{1i} = R_{2i} = 0.20$, $C_{1i} = C_{2i} = 1$, $h_1(\eta_1) = N(0, 1)$, $h_2(\eta_1) = N(0, 2)$.

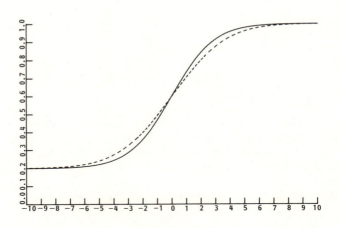

FIGURE 1. Examples of item characteristic curves generated by formula (5). Item parameters: $\alpha_{1i} = 0.46$, $\alpha_{2i} = 1$, $R_{1i} = R_{2i} = 0.20$, $C_{1i} = C_{2i} = 1$. Distribution parameters: ———, $h_1(\xi, \eta_1) = N(0, 1, 0, 1, 0)$; – – –, $h_2(\xi, \eta_1) = N(0, 1, 0, 2, 0)$.

5. APPLICATIONS

Explaining item bias in terms of psychological constructs is sometimes a difficult enterprise. It is seldom easy to interpret the bias of items that are suspected of bias by some detection procedure. Several authors acknowledge that all detection methods have their drawbacks, which may manifest themselves under specific psychometric conditions (see Berk, 1982). Therefore, wrong decisions are sometimes made that give rise to interpretation problems. Research concerning the adequacy of detection procedures under certain conditions has been carried out frequently. In this research, often, though not always, biased test data are simulated by the computer.

Two lines of research are relevant in this context. Several studies have been reported that compare the behavior of different bias detection techniques by computer in order to determine the best procedure (Merz & Grossen, 1979; Rudner, Getson, & Knight, 1980). In these studies bias often is induced by varying discrimination or difficulty item parameters of the logistic model over groups. An alternative, provided by the model presented in this paper, could be the simulation of bias by varying distribution of additional latent traits, over groups. Although, of course, item parameters in formulas (3) and (1) must still be specified, the interpretation of the simulated bias may be easier. Moreover, by applying the model for the simulation it can be investigated whether different detection procedures are equally sensitive to different types of bias caused, for instance, by different group distributions of "compensatory" or "conditional" abilities, or testwiseness.

Another important line of research, where computer-simulated test data are used, is the research of the statistical power of the detection procedure. Scheuneman (1982) asserts that detecting common features in *several* biased items may be very helpful in explaining bias. In the context of the interpretation problem, therefore, a sound detection procedure should not only classify truly unbiased items as unbiased, but also classify rather biased items as biased. This means that given the number of testees in both groups, the length of the test, and the significance level, the power of the procedure must be sufficiently large. In order to investigate the power by simulation research, some model accounting for test behavior must be specified and parameters chosen for n items such that bias of a certain effect is induced with some items. Samples of N_1 and N_2 subject parameters (or vectors) are selected at random from distributions of specified shape. A $(N_1 + N_2) \times n$ response matrix is generated and analyzed by the detection method studied. This procedure is repeated a great number of times. The proportions of the bias classifications of different items are estimations of the power of the detection-method in relation to the induced bias effect.

Very few studies of this kind have been reported. Kok (1985a, b); Kok and Mellenbergh (1985); and Lucassen and Evers (1984) describe power research of one specific bias detection procedure, the iterative logit method (van der Flier, Mellenbergh, Adèr, & Wijn, 1984). Kok applies the model reported in this chapter for generating test data, with different types of bias. He also gives definitions of bias effect size in terms of latent trait distributions.

Even if some detection procedure identifies biased items well, there may be interpretation problems, because the occurrence of item bias may be a very subtle process. Scheuneman (1982) gives a clarifying account of problems encountered, while trying to find an *a posteriori* explanation of item bias. She also describes research (Scheuneman, 1985) where very detailed hypotheses about characteristics of items are formulated. Viewing item bias as originating from different conditional group distributions of different abilities provides means for formulating the explanation problem in an accurate manner and for describing in a systematic fashion different research strategies. For instance, the main interest of a particular researcher may be either in "finding the biasing traits" (in a certain population (see e.g. Mellenbergh, 1984) or in discovering characteristics that render items "sensitive" for certain traits. Both lines of research may lead to hypotheses that can be investigated by correlational, quasiexperimental, or experimental research. Many bias explanation studies reported in the literature can be classified according to these two dimensions (see Mellenbergh, 1984).

6. CONCLUSIONS

In this chapter, a model is presented that "explains" the occurrence of noncoinciding ICCs in terms of test multidimensionality. The idea behind the model can be expressed by a simple example.

If we were to compare members of the black population with members of the white population of the same arithmetic level, of the same verbal ability level, with the same powers of concentration, and so forth, there would be no difference in item response success probability. However, the issue of fairness implies that equal success probabilities should also be observed, if we compare members of the black group with members of the white group of equal arithmetic ability only (if the test is supposed to measure arithmetic ability).

The model is used for computer simulation research. The question to be addressed next is whether techniques can be developed for further applications. For instance, information about which additional latent traits are operating could possibly be acquired by applying some form of nonlinear factor analysis for dichotomous data. Or, if other measurements of additional traits are available, certain hypotheses could be tested by comparing nonlinear binary regression models.

7. REFERENCES

Angoff, W. H. (1982). Use of difficulty and discrimination indices for detecting item bias. In R. A. Berk (Ed.), *Handbook of methods for detecting test bias* (pp. 96–116). Baltimore: Johns Hopkins University Press.

Berk, R. A. (Ed.). (1982). *Handbook of methods for detecting test bias*. Baltimore: Johns Hopkins University Press.

Flier, H. van der, Mellenbergh, G. J., Adèr, H., & Wijn, M. (1984). An iterative bias detection procedure. *Journal of Educational Measurement, 21,* 131–145.

Green, D. R. (1982). Methods used by test publishers to "debais" standardized tests: CTB/McGraw-Hill. In R. A. Berk (Ed.), *Handbook of methods for detecting test bias* (pp. 229–240). Baltimore: Johns Hopkins University Press.

Hunter, J. E. (1975). *A critical analysis of the use of item means and item test correlation to determine the presence or absence of content bias in achievement test items.* Paper presented at the National Institute of Education Conference on Test Bias, Annapolis, MD.

Kok, F. G. (1985a). *Een simulatie onderzoek naar het functioneren van de Iteratieve Logit*

Bias detektie Methode (Report). Amsterdam: Netherlands Association for the Advancement of Pure Research, University of Amsterdam.

Kok, F. G. (1985b, August). *Item bias: A special case of test multidimensionality.* Paper presented at the 13th Symposium on Latent Trait and Latent Class Models in Educational Research, Kiel, Federal Republic of Germany.

Kok, F. G., & Mellenbergh, G. J. (1985, July). *A mathematical model for item bias and a definition of bias effect size.* Paper presented at the Fourth Meeting of the Psychometric Society, Cambridge, Great Britain.

Linn, R. L., Levine, M. V., Hastings, C. N., & Wardrop, J. L. (1981). Item bias in a test of reading comprehension. *Applied Psychological Measurement, 5,* 159–173.

Lord, F. M. (1980). *Applications of item response to practical testing problems.* Hillsdale, NJ: Lawrence Erlbaum Associates.

Lord, F. M., & Novick, M. R. (1968). *Statistical theories of mental test scores.* Reading, MA: Addison-Wesley.

Lucassen, W., & Evers, A. (1984). *Oorzaken en Gevolgen van Sexe-Partijdigheid in de Differentiële Aanleg-Testserie, DAT'83.* Paper presented at the Congress of the Netherlands Institute of Psychology, Ede, The Netherlands.

Marasquilo, L. A., & Slaughter, R. E. (1981). Statistical procedures for identifying possible sources of item based on χ^2 statistics. *Journal of Educational Measurement, 18*(4), 229–248.

Mellenbergh, G. J. (1984, December). *Finding the biasing trait(s).* Paper presented at the conference: Advances in Measuring Cognition and Motivation, Athens, Greece.

Merz, W. R., & Grossen, N. E. (1979). *An empirical investigation of six methods for examining test item bias.* (Final Report No. NIE-6-78-0067). Sacramento: California State University.

Muthén, B., & Lehman, J. (1985). Multiple group IRT modeling: Applications to item bias analysis. *Journal of Educational Statistics, 10*(2), 133–142.

Rudner, L. M., Getson, P. R., & Knight, D. L. (1980). A Monte Carlo comparison of seven biased item detection techniques. *Journal of Educational Measurement, 17,* 1–11.

Scheuneman, J. D. (1982). *A posteriori* analyses of biased items. In R. A. Berk (Ed.), *Handbook of methods for detecting test bias* (pp. 180–198). Baltimore: Johns Hopkins University Press.

Scheuneman, J. D. (1985). *Exploration of causes of bias in test items* (Report GRE No. 18–21). Princeton, NJ: Educational Testing Service.

Veale, J. R., & Foremen, D. I. (1983). Assessing cultural bias using fail response data: Cultural variation. *Journal of Educational Measurement, 20,* 249–257.

CHAPTER TWELVE

On a Rasch-Model-Based Test for Noncomputerized Adaptive Testing

KLAUS D. KUBINGER

1. INTRODUCTION

There is no doubt that the development of latent trait theory and models has made adaptive testing almost an everyday challenge for psychometricians (cf. the well-known review of Weiss, 1982, or Wild, 1986). However, efforts towards adaptive testing are, primarily, restricted to computerized administration. Conventional paper-and-pencil tests where no computer is necessary for administration are rare, particularly published tests available to any practitioner. For this reason I give an example of how such a test may be calibrated; the example concerns a revised issue of the German WISC which is a prototype of a test battery used worldwide. Of course, suiting it to computerized adaptive testing would most likely mean forfeiting a chief purpose of Wechsler's concept, that is, to include the testee's interaction with the examiner. Nonetheless, an improvement in this test battery's measurement precision and/or a reduction in the test length by means of adaptive testing would be a welcome move to innumerable practitioners. In this chapter the following topics are to be discussed: item generation and administration, model checks and cross-validation, error of estimation, learning, and last but not least, handling by practitioners.

KLAUS D. KUBINGER • Institute of Psychology, University of Vienna, A-1010 Vienna, Austria.

2. THE ADAPTIVE INTELLIGENCE DIAGNOSTICUM (AID)

The AID (Kubinger & Wurst, 1985) is based on the German WISC which has contradicted psychometric presuppositions in many respects (cf. Kubinger, 1983). The aim was, first, to eliminate WISC's shortcomings and, second, to develop an adaptive test issue. The latter, for instance, was desired because originally all the same items were conceived for all children without regard to age. Besides this, the AID consists of corresponding subtests; it is conceptualized for children from 6 to 16 years old. In the following we shall deal mainly with the first subtest, that is Everday Knowledge corresponding to Information.

3. THE AIMED-FOR BRANCHED TESTING DESIGN

In the first instance it was decided to base item calibration on the Rasch model for reasons of simplicity. Parameter estimation, in contrast to Birnbaum's models, causes no problem and the same is true for the scoring rules to be applied by practitioners. In the second instance it was decided to aim for an administration design according to Figure 1: There are different subsets of five items each, upleveled from left to right. Depending on the age, every testee is given an *a priori* fixed subset; afterward, however, the subset to be administered depends solely on the testee's achievement level. If the child solves one item at the most, a subset one level below is recommended; if at least four items are solved

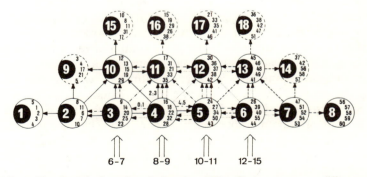

FIGURE 1. The branched testing design for AID. Circles represent different subsets up-leveled from left to right. Each subset contains five items, the number of each corresponding to the rank of difficulty. The age of the testee determines the starting point. Because branched testing terminates after the third subset, dashed-line subsets consist of items of solid-line subsets.

then a subset one level higher up is advised, and only if two or three items are solved is the subset to administer at an equivalent level. This design guarantees that every testee is tested by almost optimal items; in the long run it is not likely that the items administered will be too easy or too difficult, but it is most likely that many high-information items will be used. In other words, the standard error of estimation (SEE) comes close to the ideal minimum that would be achieved by tailored testing.

Standard administration was intended to be terminated after the third subset. As a consequence 60 different homogeneous items would satisfy this design if some of them were deliberately included in more than one subset.

From the very beginning it was, nevertheless, part of the concept that every examiner might abandon the adaptive testing idea and choose some proper looking subset combination—proper with respect to a particular child or proper with respect to all children of his clientele. In addition, it was conceptualized to offer short forms as well as parallel forms. As an instrument for screening, the former might be selected again at the examiner's discretion or in accordance with the standard administration by termination after the second subset. And because the item pool consists of many more items than any WISC subtest, it is easy to select a relatively optimal individual parallel form in case a second test administration occurs.

4. ITEM CALIBRATION

With the help of practitioners, an item pool of 100 subtest items resulted, which proved to have content validity. Because it is impossible to examine every subject of a sample against all 100 items, the pool had to be subgrouped, but the subsets still had to be linked—otherwise no parameter estimation would have been possible. For this, the schedule given in Figure 2 was mainly applied. Thirteen subsets with 20 items each were administered conventionally to 1068 children of both sexes between the ages of 6 and 16, all from Austria, the Federal Republic of Germany, or Switzerland. Of course, the 100 items were ranked intuitively in advance according to their difficulties. An additional 232 children were tested with another schedule.

Item analyses referred to several "internal" model checks as well as to three external ones: While the former contrast two polarized subject score groups (high *versus* low raw score) with respect to their item parameter estimations, the latter polarize subject groups according to

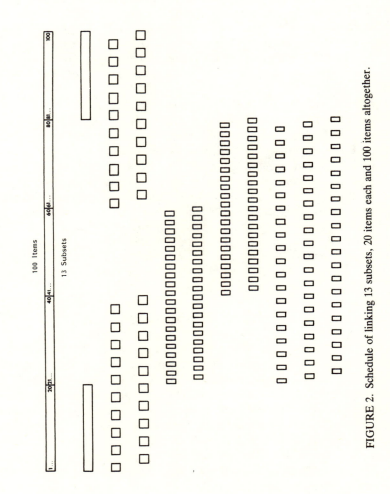

FIGURE 2. Schedule of linking 13 subsets, 20 items each and 100 items altogether.

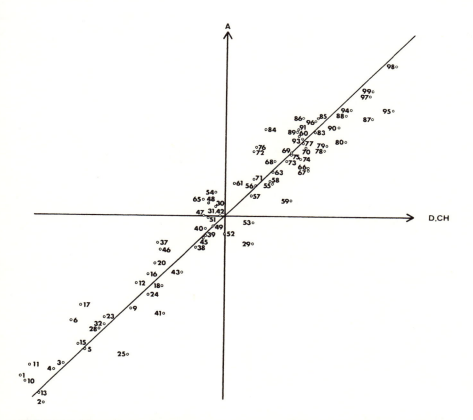

FIGURE 3. Graphic model check with respect to nationality: Likelihood Ratio Test is significant.

sex, age, and nationality.[1] As was expected, most items that failed to fit the Rasch model did so with respect to nationality ($\alpha = .01$). Compare the graphical model check in Figure 3, where the item difficulty parameters are compared as estimated for Austrian children on the one hand and for German and Swiss children on the other hand. While in Figure 3 all estimatable items are included (Likelihood Ratio Test $\chi^2 = 264.68$, df = 79), Figure 4 demonstrates a nonsignificant model check after stepwise deletion of nonconforming items ($\chi^2 = 74.18$, df = 50). As there are some more items that do not explicitly contradict

---

[1] An unpublished computer program RML100 by Foreman, Kubinger, Wild, and Gittler was used.

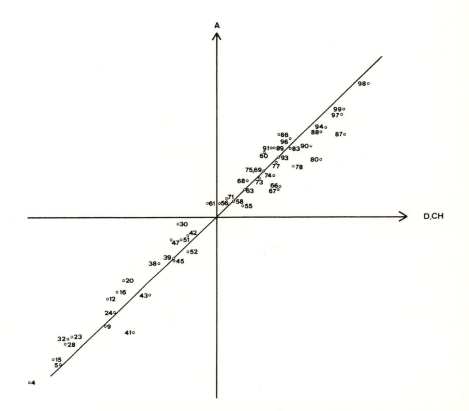

FIGURE 4. Graphic model check with respect to nationality: Likelihood Ratio Test is not significant.

the Rasch model but are simply not analyzable because in the given sample they were generally solved or unsolved, the number of remaining items amounts to 60. This is, in fact, the number aimed for, so branched testing according to Figure 1 is, in principle, possible.

We have to recognize that a nonsignificant model check after stepwise deletion of nonconforming items always means an increased type I error and never guarantees a homogeneous test. To surmont these obstacles, some cross-validation is absolutely necessary. Such a cross-validation was done at the next stage, because in order to standardize the test battery a second sample was tested, but this time individualized as provided for by the branched testing design. This sample included 1460 children. As Figure 5 shows, the difficulty parameters of any item in

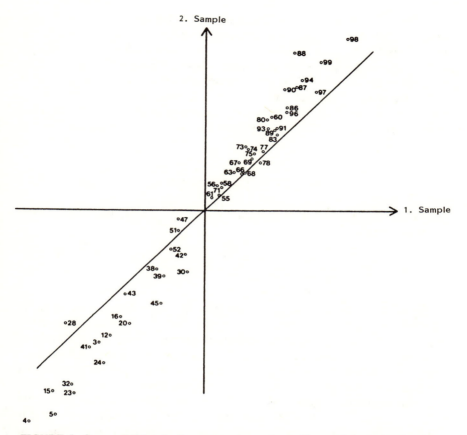

FIGURE 5. Cross validation for item parameters estimated by two independent samples.

question vary negligibly when estimated in two independent samples.[2] We may, therefore, state that these items constitute a Rasch-homogeneous test.

There is an interesting empirical fact: The optimal straight line that might represent the points in Figure 5 is not accurate at 45°, but has a steeper slope. This is true for all the subtests, that is, the item difficulties' distribution is slightly more extensive when the items are administered adaptively and not conventionally. The reason might be that in the latter

[2] No Likelihood Ratio Test was done because of the problem of CP memory: The program RML100, which applies to 100 items and 26 different subsets of items, comes close to maximum capacity, but analysis of both samples at once would mean almost twice as many subsets.

case items sometimes occur that are out of the range of an individual ability parameter but are, nevertheless, solved or not solved by chance; as a consequence their parameter estimations are less extreme. Kubinger (1983) found a similar phenomenon when he analyzed data with or without regard to the given discontinue rules that should be applied in case of too many failures by the testee; if items not administered to a testee are scored as "wrong" then the distribution of item difficulties is a little more extensive than if they are left out of computation for this testee.

5. THE EVALUATION OF "SEE"

The standard error of estimation for the Rasch model is widely known as being

$$\text{SEE } (\xi_v) = \left[\sum_{i=a_1(v)}^{a_k(v)} \frac{\exp(\xi_v - \sigma_i)}{1 + \exp(\xi_v - \sigma_i)} \frac{1}{1 + \exp(\xi_v - \sigma_i)} \right]^{-1/2} \quad (1)$$

ξ_v is an arbitrary ability parameter and the σ_i are the item difficulty parameters—say we consider subject v to whom k items $a_1(v), a_2(v), \ldots, a_k(v)$ are administered. Given appropriate estimates of σ we may, therefore, study the test's reliability along the ξ continuum. Figure 6 portrays this graphically. As concerns AID, three representative paths of the standard form according to Figure 1 were selected, as well as the corresponding short and parallel forms. Although the respective subtest of the German WISC has been proved inadequate according to the Rasch model and, furthermore, parameter estimations are based on another sample (cf. Kubinger, 1983), we may use it for one more comparison. For a better illustration, just those ξ were taken into account that represent maximum likelihood estimations of potential raw scores.

First of all, the AID places more importance than the German WISC on the range of ability to which the test applies, because the same is true for the item difficulty parameters: Although in the German WISC they are ranged between −6.02 and 6.51, in the AID they are ranged between −8.7 and 7.4, i.e., the AID fits better for children characterized by extreme ability. In particular, even where both test batteries apply, the AID offers a much smaller SEE. No heed should be paid to a medium ability where the opposite is true: There are some other paths of branched testing for AID that are not represented but come closer to conventional testing by German WISC; and the latter is superior only

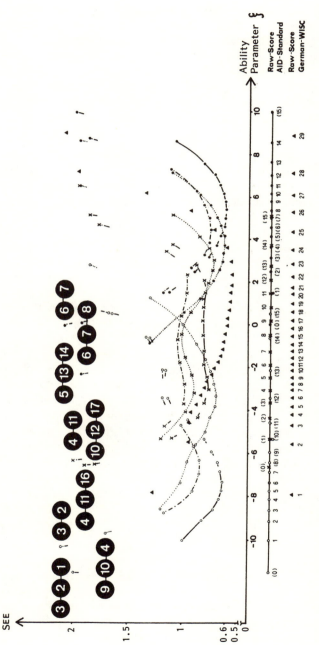

FIGURE 6. Standard error of estimation depending on ability and raw score, respectively. To compare 30 conventionally administered items of the German WISC, 15 and 10 adaptively administered items of AID, and 15 corresponding parallel form items. ▲▲▲ = German WISC; ——— = AID—standard form; ··· = AID—parallel form; –·–·– = AID—short form.

when at least 16 items are administered, including those five necessary because of discontinue rules—it should be borne in mind that AID testing never means administration of more than 15 items.

Second, SEE is evidently enhanced if either a short form or a parallel form is used. For the latter, the subsets selected are less optimal than in the standard form. However, even the adaptive short form results in less SEE than the conventional long form of the German WISC when dealing with extreme ability.

6. THE PROBLEM OF LEARNING

Of course, adaptive testing can only be taken seriously when there are no learning effects at all. If a subject's ability were to benefit from the number of items he had already tried to solve or even from the items themselves, the individual difficulties of later items would change. As a consequence no fair comparison of achievement is possible for subjects to whom the same subset of items is administered, but where the sequences differ; this is because the same items would not present the same difficulties. On the other hand, such learning effects would just be absorbed in a fair way into the difficulty parameters and pose no problem if all items were administered conventionally, that is, in the same sequence for all subjects.

As a matter of fact, special learning effects were discovered as concerns AID's Block Design. When, for psychometric analyses, all items were taken into account, no Rasch homogeneous test resulted after stepwise deletion of nonconform ones. However, when those two individual items placed at the very beginning of a child's examination were not taken into account, the subtest soon fitted the Rasch model. So psychometric analyses confirmed the intuitive suspicion that warming-up effects occur. Therefore, at the end of calibration it was decided to introduce two obligatory warming-up items that are not to be scored. In a forthcoming paper Kubinger and Wild (in preparation) find by means of a better-founded reanalysis that no other learning effects are established. For this Fischer's (1972) linear logistic test model was used in order to give full regard to the items' sequence.

7. THE OTHER SUBTESTS OF AID

Disregarding Digit Span and Coding, all other subtests of the AID were also conceptualized for branched testing. Actually, a Rasch-homogeneous test always resulted. Just because of the extra time the

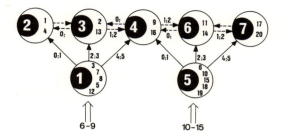

FIGURE 7. An alternative branched testing design for AID. For the purpose of reducing SEE, branched testing for Picture Arrangement and Block Design may be prolonged by using the dashed-line arrows.

items' solutions require, the design of Figure 1 was simplified for the performance tests. This means there are fewer items to administer. And every subtest, disregarding Object Assembly, is to be scored according to Rasch's dichotomous model.

At first glance the WISC seems highly superior to AID as concerns the number of graduable achievements and raw scores, respectively, because there most of the items in the performance tests are to be scored multicategorically. However, adaptive testing substitutes a lot of the information we lose by dichotomous scoring. Moreover, the AID offers extended forms that are to be used facultatively, that is, branched testing may be prolonged by some items (cf. Figure 7). Nevertheless, the time required for administration does not surpass the German WISC because of the shortened time limits. It should be remembered that multicategorical scoring poses a serious psychometric problem, particularly the confusion of speed and power. There is, in fact, empirical evidence that not all presuppositions are fulfilled as concerns German WISC's Picture Arrangement and Block Design (cf. Kubinger, 1983).

The objection of loss of information does not hold for Object Assembly. This subtest was calibrated for the AID according to Rasch's multicategorical unidimensional model (cf. Andersen, 1980), the items being scored three-categorically. Hence, the number of graduable achievements exceeds the German WISC. Though not indicated, the item pool proved to fit the model if correct responses were scored either as "quick" or "slow," whereas it did not if partially correct responses were partially credited. As a matter of fact, a slow but correct response has turned out to weigh half as much as a quick correct response. In the end, multicategorical scoring does not prevent branched testing.

8-9

1. Name a conifer. /Fir, pine, cypress, larch, juniper, cedar, yew.

2. During which month do we celebrate Christmas? December.

3. Which rodent builds dams in water from tree trunks? Beaver.

4

4. Who was Napoleon I? French Marshal, Emperor.

5. Approximately how heavy is one liter of water?/ 1 kg.

	6-7	8-9	10-11	12-15
0;1	⑩	❾	❽	–
2;3	⑪	⑪	⑪	–
4;5	❻	❻	⑫	–

FIGURE 8. Extract from AID's manual. According to a testee's raw score the examiner is recommended to choose the next subset.

8. THE HANDLING OF AID

Of course, noncomputerized adaptive testing places exacting demands upon the examiner. However, the 40 or so psychologists and students of psychology appointed to test a total of 2778 children turned out to have no handling problems after the first two or three children each. The reason is that the examiner does not in fact have to work through the design as shown in Figure 1, but a programmed instruction directs him in accordance with this same design (see as an illustration Figure 8). After the appropriate subset of items has been administered, the examiner is directed to the next subset according to the number of items the testee has solved. At most there is a problem for practitioners accustomed to the WISC because now every item is to be scored at once, yet the examiner has no time at his disposal to reflect upon response. Therefore it was necessary to give unequivocal responses in the AID directions for scoring. This challenge was met by abandoning multi-categorical scoring as far as possible.

There is evidence that administration of the standard form requires no more than 60 minutes. Short forms may last only 20 minutes.

We have also realized that branched testing enhances testees' motivation in examination. Compared to the WISC, the testees are neither frustrated nor bored as a consequence of too difficult or too easy items. Rather, the items' level is adequate to the testee's ability level so that correct and incorrect responses alternate. Thus items always occur that the testee solves, although he has failed some previous items. It is of importance to note that because of discontinue rules almost every child finishes almost every subtest of WISC after three to five failures.

Although scoring already takes place during administration, AID needs some computation work afterward. First of all, achievements within different subsets have to be made comparable to each other. Therefore, the ability parameters' estimations are tabulated for every raw score in many combinations of subsets. However, in order to draw comparisons with the population of the same age, a second table is to be used; for this, T scores are tabulated for every age and many potential ability parameters.

9. CONCLUSIONS

There is, after all, impressive evidence that for commonplace problems noncomputerized adaptive testing works; and there is evidence as well for how it works. That might supply a reason for applying AID, but more importantly it should encourage psychometricians to develop noncomputerized adaptive tests for fields other than those based on the experiences reported here. As illustrated, this would be much harder, of course, than conceptualizing some conventional test, but we remain confident that this is the right direction towards a scheme for fair testing.

10. REFERENCES

Andersen, E. B. (1980). *Discrete statistical models with social science applications.* Amsterdam: North-Holland.

Fischer, G. H. (1972). Conditional maximum-likelihood estimation of item parameter for a linear logistic test-model (Research Bulletin No. 9). Vienna: University Vienna, Institute of Psychology.

Kubinger, K. D. (Ed.). (1983). *Der HAWIK—Möglichkeiten und Grenzen seiner Anwendung.* Weinheim: Beltz.

Kubinger, K. D., & Wild, B. (in preparation). *Kognitive Prozesse und Lerneffekte beim Lösen von Mosaik-Testaufgaben.*

Kubinger, K. D., & Wurst, E. (1985). *Adaptives Intelligenz Diagnostikum (AID).* Weinheim: Beltz.

Weiss, D. J. (1982). Improving measurement quality and efficiency with adaptive testing. *Applied Psychological Measurement, 6,* 473–492.

Wild, B. (1986). *Der Einsatz adaptiver Teststrategien in der Fähigkeitsmessung.* Doctoral dissertation, University of Vienna.

CHAPTER THIRTEEN

Systematizing the Item Content in Test Design

FONS J. R. van de VIJVER

1. INTRODUCTION

Traditionally, constructing a test frequently amounts to writing a number of items, administering these items to a large sample of subjects, and—on the basis of these data—selecting a set of apparently appropriate items. This procedure cannot be expected to yield instruments that will satisfy the assumptions of item response theory (for a recent overview of these the reader is referred to Hambleton & Swaminathan, 1985). The weak spot in this procedure is the item writing. Items frequently differ in many ways, some of them of interest to the investigator, but some of them hardly specifiable.

One consequence of this approach is the often-reported lack of cross-sample stability of item indices, for example, item discrimination indices or fit statistics in item response theory. An area in which this phenomenon has been amply demonstrated is the research on cultural bias. In this paradigm a test typically is administered to at least two samples belonging to different cultures. A major concern in the data analysis is the detection of items with differential statistics across cultural groups. It is common to find a lack of cross-sample replicability, and the psychological reasons for the presence of item bias in a particular subset of test items are often unclear (cf. van de Vijver, 1984). Despite the fairly large number of relevant empirical studies carried out in this paradigm,

FONS J. R. van de VIJVER • Department of Psychology, Tilburg University, 5000 LE Tilburg, The Netherlands.

this research has not led to a cumulative body of knowledge about how to design instruments suitable for intergroup comparisons.

In the present paper a different approach is adopted in which the content of the test items is systematized to a considerable degree. Central in our test design is Guttman's facet theory (a recent introduction can be found in Canter, 1985).

As an example, suppose a researcher wants to compose an arithmetic test. First a number of "constituent elements" are sought. The first element, or facet, of each item consists of the kind of operation required, for example, addition, subtraction, multiplication, and division. In terms of facet designs this is called a facet with four levels. Also, suppose that the research wants to use numbers with one, two, and three digits in the test. This constitutes a second facet, with three levels. By forming the Cartesian product—called crossing in an analysis of variance—twelve kinds of items can be composed (four levels of the first facet times three levels of the second facet).

The use of facet designs offers some important advantages. First, tests constructed by means of facet designs have a meaningful internal structure. Items are explicitly defined in terms of the facet design underlying the test.

In many instances it is possible to order facet levels according to their difficulty level prior to the testing. These expected rank orders can be compared with the actual outcomes, which supplies a check of the construct validity of the instrument.

Another advantage of facet designs is their possibility for formulating conclusions about meaningful subsets of items. If groups of subjects differ with respect to the number of task relevant concepts, that is, facet levels, they have mastered, it is possible to restrict comparisons to those levels mastered by all groups. These differential skill patterns are not uncommon in developmental or cross-cultural research. Furthermore, by means of latent trait theory, more particularly the Linear Logistic Test Model (Fischer, 1974, 1983), the difficulty of the various levels of each facet can be estimated and mutually compared, and the fit of the model can be evaluated.

It has been shown under various conditions that the Linear Logistic Test Model can yield a good approximation of item difficulties (Chapter 18 of Fischer's 1974 book contains many applications of the model). Moreover, Fischer (1976) has described a procedure to measure change by means of this model, while Spada (1977) has used the model to analyze learning behavior and curriculum evaluation.

In the remainder of this chapter two examples of tests constructed on the basis of facet designs will be described. The data to be presented have been gathered in a cross-cultural study of inductive reasoning.

2. METHOD

2.1. SUBJECTS

In three countries, The Netherlands, Turkey, and Zambia, pupils from four subsequent grades were tested. In a pilot study the lowest grade to be included was determined for each country, being the lowest grade in which a substantial proportion of the children has a score above chance level. In Zambia the lowest grade tested was the sixth grade, while in Turkey and The Netherlands the fifth grade was chosen. In each grade 95 pupils were tested with approximately an equal number of girls and boys. The Zambian pupils, tested in the English language, were recruited from primary schools (Grades 6 and 7) and secondary schools (Grades 8 and 9) in Lusaka and the rural surroundings (up to 150 kilometres from Lusaka). In Turkey pupils were taken from the highest (5) grade of primary schools and the three grades of middle schools in Istanbul and rural areas up to 100 kilometres away from Istanbul. In The Netherlands pupils from a number of urban and rural schools in the southern provinces were administered the tests. In each country the tests were administered by local assistants who had prior experience in testing.

2.2. TESTS

Two unspeeded tests for inductive reasoning, a verbal and a figural test, were administered.

2.2.1. Letter Test

The former, called the *Letter Test,* is derived from the Letter Sets Test in the ETS-Kit of Factor-Referenced Tests (Ekstrom, French, & Harman, 1976). This test presumably reflects inductive reasoning in a symbolic medium. Each item in the Letter Test consists of five groups of six letters. Four out of these five groups have been generated according to some rule, while a fifth one does not follow this rule. The subject has to mark this latter group. In the test of 45 items the following five item-generating rules (and only these) are used:

1. Each group of letters has the same number of vowels, for example, AEVZBR AIGVRS AVRGTK EUSRZQ UENBZR.
2. Each group of letters has an equal number of identical letters, which are the same across groups, for example, VVVVRG QZHGSL VTVVVB RFVVVV RFVVVV.
3. Each group of letters has an equal number of identical letters,

which are not the same across groups, for example, RGVSBD
CCCFHR WSWWGY GHKZZZ LMLLVS.

4. Each group of letters has a number of letters that appear the
same (i.e., 1, 2, 3, or 4) number of positions *after* each other in
the alphabet, for example, BCDZRG QRSWVB KLMGSV
FGHZDD VWHNKP.

5. Each group of letters has a number of letters that appear the
same (i.e., 1, 2, 3, or 4) number of positions *before* each other in
the alphabet, for example, SVGTRP ZXVKBP NLJBWX
FDBHVQ GRVTSZ.

In addition to this first facet a second one was introduced, namely,
the number of letters to which the rule applies. This number varies from
one to six. In the first example (about the vowels) the rule applies to two
letters, while in the second example the rule refers to four letters.

The fourth and fifth rules refer to a difference in position in the
alphabet. These two rules define a final facet, namely, the distance (of
one, two, three, or four positions) between letters in the alphabet. The
example of rule five refers to a difference of two positions. The English
and Dutch version of the test and its facet design appear in Figure 1.

As the Turkish alphabet on the one hand and the Dutch and English
alphabet on the other hand are not completely identical, a somewhat
different test had to be composed for the Turkish pupils. It is fairly
obvious that, given a particular facet design, parallel tests or adaptations
such as needed here are easily composed, as each item is uniquely
defined by its location in the design matrix.

Although Figure 1 may suggest otherwise, the actual number of
columns in the total design matrix is 15. Six columns of the design matrix
refer to the number of letters to which the rule applies; the following five
columns refer to the five item-generating rules; and the final four columns
only apply when the fourth or fifth item-generating rule has been used.
These latter columns represent the difference in number of positions in
the alphabet. When this full design matrix would be used in a Linear
Logistic Test Model analysis the problem would arise that it is not of full
rank. Fischer (1983) has demonstrated that the design matrix should be
of full rank in order to allow for unique estimates of facet difficulties. The
rank of the present design matrix will be the number of columns (i.e., the
number of facet levels) minus the number of facets, being $15 - 3$.

On the basis of the facet design a number of hypotheses about the
expected rank order of item difficulties can be postulated, which will
provide a check on the construct validity of the instrument. Items will be
easier when they deal with a larger number of letters. With respect to the

FIGURE 1. Items and design matrix of the Letter Test.

	Items				Number of letters	Rule	Number of positions
1. GGGGGG	GGGGGG	GGGGGG	GGGGGP	GGGGGG	6	2	
2. XXPXXX	ZZRZZZ	CCCCCH	YDDDDD	FFFFPQ	5	3	
3. QIAEOU	WEIAVC	EIOUAZ	OUIEPA	WUEIOA	5	1	
4. HHHHHI	PPPPPP	RRRRRR	BBBBBB	JJJJJJ	6	3	
5. BCDEFQ	KLVMNO	RSUVWZ	FGHMIJ	TUVCWX	5	4	1
6. MLKJIH	GFEDCB	UTSRQP	ONMLKH	XWVUTS	6	5	1
7. BXDDDD	YDDDDD	DRDDDD	DDDSDD	DDYDDD	5	2	
8. FGHIJK	RSTUVW	ABCDFG	JKLMNO	PQRSTU	6	4	1
9. AAIOUU	IEEOOA	IIUUEA	AAUEEI	AEIOUZ	6	1	
10. PFEDCB	BWVUSR	KJIHGR	WVUQTS	JIHGLF	5	5	1
11. EAOIQU	WAOIEU	IOEXAU	AEPQIO	EAUVOI	5	1	
12. HHHHHI	HHHHHH	HHHHHH	HHHHHH	HHHHHH	6	2	
13. QQQQRQ	VSVVVV	TTTETT	BBABBB	DDYYDD	5	3	
14. AEEAIO	IIEESU	OOUEEI	IIAAEE	EEOOUI	6	1	
15. NNNNON	NNQNNN	RNNNNS	NNZNNN	YNNNNN	5	2	
16. RBBBBZ	XBCBBB	BBCBCD	RSBBBB	BBBBZZ	4	2	
17. GHHHHZ	KVKKKM	MMMPMT	RRRSTT	VVVVXY	4	3	
18. BCQDPZ	VWKXSZ	KLPPMM	QRLJVB	FGRHTY	3	4	1
19. VAEUSZ	EEOIPR	EIOUSV	UUEABG	OIEAQW	4	1	
20. PLRRWR	XVVVBD	NNYYTX	KKKQWR	DBDDFZ	3	3	
21. CXEGIK	RTKVXZ	KMOQBS	NPQRWZ	HJPLNP	5	4	2
22. OQTRPZ	GAEZRW	EDUXGK	IBWALV	GKAESW	2	1	
23. WMLRKP	DWVULG	SHGBFB	OUNMXK	BAFKRV	3	5	1
24. RRTURW	RRXSRA	WRXVRU	KRRBRK	RRDRPV	3	2	
25. LJHFDB	MKIGEC	ZXVTRP	PNLJHF	ECAACE	6	5	2
26. HIJTKZ	ACEGIK	PQBFRS	DJKLMV	RFSTUY	4	4	1
27. GIKMOQ	LNPRTV	PRTVXZ	ACEGIK	KMORVZ	6	4	2
28. FZIUEQ	RAIOXX	UORXDB	UECALW	PKEBIO	3	1	
29. DRPWQH	YBICJL	KSLBGS	PWNKRD	RAPBXZ	2	4	1
30. PNDLHJ	SQOMKX	JHFEDB	ZXVTPR	STQMJG	5	5	2
31. RZBJFN	CGUZZG	HLPTDL	QNRVZB	EIMQBB	4	4	4
32. FAGHAR	ZVAWTA	AQMAGV	SAKVBR	DAMFAH	2	2	
33. QMIEAL	GERLYQ	WVSOKG	ZVRNMJ	TPLQHD	5	5	4
34. XROPVF	NVEWGN	HLPKVD	RCDIZT	PBGAWT	1	1	
35. DKLQWZ	KRDTEH	VHRNCX	FBHNSW	MBFOWK	2	4	2
36. KHEBVR	WTQNLL	JGDAWZ	UROLBK	XPLGFF	4	5	3
37. WBMLVR	DTFHSY	DCQNLR	HRLZSD	YBLWSP	1	2	
38. XDBKRE	TNLLRK	VEGVSB	PFCCHL	OZAGZQ	2	3	
39. KVRSWU	PSDVYB	DGJTMN	FIXLRO	QTPWDZ	4	4	3
40. TRNJFV	XTPLDS	MIEAQT	VKOKGC	PFDVXB	4	5	4
41. PVXFDB	SVRYYD	QWZLJH	GHPANL	RZWWXV	3	5	2
42. VRNJFB	ZTMGDB	XTPLHD	UQMIEA	WSOKGC	6	5	4
43. AYXKWV	BNMLPK	BCDFGH	IHGFZR	YUWTSR	4	5	1
44. FPNMXY	UDMCFP	PSRDKN	KVLLWC	GYXVJB	2	5	1
45. YVPHDA	CRBFOO	WGKGSV	KLRVJQ	OSFDPZ	2	4	4

fourth and fifth item-generating rule it may be expected that items will be more difficult when the distance of the letters is large. Moreover, items dealing with identical letters will be easier than items about a difference in position in the alphabet. Finally, items about identical letters in a group that are equal across groups will be easier than items about identical letters in a group that are not the same across periods.

2.2.2. Figure Test

The second test, called the *Figure Test,* is a figural completion test, presumably reflecting inductive reasoning in a figural medium. Each of the 30 items consists of five rows of twelve figures. The first eight figures are the same for all five rows, while the final four figures differ across rows. One of the five rows has been entirely constructed according to some rule. The subject has to mark this row.

Each figure possesses a minimum of one and a maximum of six elements. A figure consists of either a square or a circle. A caret, an arrow, a dot, a dash, and a bow may or may not form part of a figure. The position of each of these elements, if present, in the figure is always the same (see Figure 2 for an illustration).

The following facets are included in the test (in the remainder of the text the italicized headings will be used to refer to the facets):

1. *Number of Figures in a Period.* The number of figures in a period, that is, the number of circles or squares following each other, may be two, three, or four. (Periods are formed by concatenating either subsequent circles or squares, but a period never contains both.) In the first item of Figure 2 there are four figures and in the second item there are three figures in a period.

2. *Number of Aspects Varied.* The number of elements in which subsequent figures in a period differ may be one, two, or three. In the first item of Figure 2 subsequent figures in a period differ in one element and in the second item in two elements.

3. *Rule.* Three different item-generating rules (and only these) have been used:

- One or more elements are added to subsequent figures in a period (cf. the first example of Figure 2).
- One or more elements are subtracted from subsequent figures in a period.
- Alternating, one or more elements are added to and subtracted from subsequent figures in a period (cf. the second example of Figure 2).

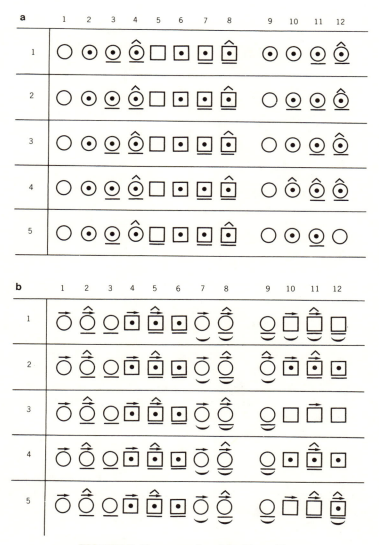

FIGURE 2. Two examples of the Figure Test.

4. *Variation across Periods.* The variation across periods, that is, the elements added or subtracted in successive periods, may be the same or differ. In both items of Figure 2 there is a constant variation.

5. *Periods (Do Not) Repeat Each Other.* Successive periods of items may or may not repeat each other that is, the first figures of each period

FIGURE 3. Facet design matrix of the Figure Test.

	Period	Variation	One aspect varied			Two aspects varied			Three aspects varied		
			Rule 1	Rule 2	Rule 3	Rule 1	Rule 2	Rule 3	Rule 1	Rule 2	Rule 3
Groups repeat each other	2	constant	1	2		5	9		6	14	
		variable									
	3	constant	8		15	7	12	11			
		variable									
	4	constant	3	4	13						10
		variable									
Groups do not repeat each other	2	constant	28							19	
		variable		24		30		26			
	3	constant		22			17				18
		variable	21					25			
	4	constant	16	29				20			
		variable			23						27

are identical apart from circles and squares. In the first example there is a repetition, while this is not the case in the second example of Figure 2.

The facet design matrix of the test is presented in Figure 3. Because it would take too many items to cover each cell of the design matrix, a selection of 30 items was made. The five facets mentioned amount to a design matrix of 13 columns, three columns each for the first three facets and two columns each for the final two facets. This design is of deficient rank; again, the rank is the number of facet levels minus the number of facets, in this case being $13 - 5$.

As for the previous test, a number of hypotheses can be postulated about the difficulty order of the facets. Items will be more difficult when they deal with a smaller number of figures in a period. Items will be easier when a larger number of aspects is varied within a period. Items will be easier when the periods repeat each other (the fifth facet) and when there is a constant variation across periods (the fourth facet). Finally, it is expected that the third item-generating rule will be more difficult than the first two rules.

The administration of both tests started with an elaborate introduction, in which a demonstration of each of the item-generating rules was given. After this instruction the pupils were asked to answer exercise items, which, again, covered all the rules. After this lengthy instruction the pupils were asked to answer the actual test items.

3. RESULTS AND DISCUSSION

3.1. GENERAL

The raw test scores are summarized in Table 1. The average performance increases with age while the standard deviations remain fairly constant. It is remarkable that the reliabilities of the Figure Test are somewhat higher than those of the Letter Test, even though the latter test contains 15 items more. This may have been caused by the fact that, in comparison with the Letter Test, the performance of the Figure Test had more an "all-or-none" character; even in the youngest group there are a number of items in the Letter Test with high p values, while this does not hold for the Figure Test.

3.2. RASCH MODEL

The *pièce de résistance* of the data analysis involves the question to what extent the theoretical structure of the tests (in terms of the facet designs) can be retrieved in each culture.

As a first step, Rasch item parameters were estimated in each sample and correlated across the samples. These correlations are given in Table 2. The correlations between Rasch item estimates were high for both

TABLE 1
Means, Standard Deviations (*SD*), and Reliabilities (Rel.) of the Tests

Figure Test (30 items)											
The Netherlands				Turkey				Zambia			
Grade	Mean	*SD*	Rel.	Grade	Mean	*SD*	Rel.	Grade	Mean	*SD*	Rel.
5	19.6	5.4	.83	5	15.2	6.6	.88	6	11.9	7.0	.89
6	22.2	5.1	.83	6	14.1	5.2	.77	7	16.3	7.5	.91
7	21.8	5.7	.87	7	21.2	4.8	.80	8	16.8	8.8	.94
8	24.9	4.2	.80	8	19.9	5.6	.85	9	17.4	7.2	.91
Letter Test (45 items)											
The Netherlands				Turkey				Zambia			
Grade	Mean	*SD*	Rel.	Grade	Mean	*SD*	Rel.	Grade	Mean	*SD*	Rel.
5	27.1	4.7	.69	5	22.5	6.2	.81	6	22.8	5.8	.79
6	28.6	5.5	.78	6	23.5	5.2	.73	7	25.8	6.0	.80
7	32.1	6.4	.85	7	30.0	4.7	.73	8	25.9	5.3	.75
8	34.9	4.2	.72	8	30.0	6.0	.78	9	27.3	6.4	.83

TABLE 2
Correlations between the Rasch Estimates

	Dutch[a]	Turkish	Zambian
Dutch	—	.90	.85
Turkish	.95	—	.87
Zambian	.96	.96	—

[a] Figure Test above diagonal; Letter Test below diagonal.

tests, ranging from .85 to .96, with the Letter Test showing somewhat higher values than the Figure Test.

Thereafter, the fit of the Rasch model was investigated by means of Andersen's (1973) test, in which the likelihood of the data matrix of the total sample ($N = 1140$) is compared with the likelihood of the data matrix in the three cultural groups. For both tests significant values for the likelihood ratio (LR) test were found (Figure Test: LR = 635.8, df = 58, $p < .001$; Letter Test: LR = 705.6, df = 88, $p < .001$), indicating that the Rasch model in a strict sense does not hold for these data. Additonally, the fit of the Rasch model was investigated in the three cultural groups separately. An LR test was computed for each cultural group in which the likelihood of the data matrix of the total sample ($N = 380$) was compared with the likelihood of the data in a group scoring below or at the median ($N = 190$) and a group scoring at or above the median ($N = 190$). For the Letter Test the LR test was significant in each group (the Netherlands: LR = 148.1, df = 44, $p < .001$; Turkey: LR = 232.0, df = 44, $p < .001$; Zambia: LR = 233.1, df = 44, $p < .001$). The same procedure was applied to the data of the Figure Test. The LR test yielded significant values in the Dutch and the Zambian sample (LR = 93.1, df = 29, $p < .001$ and LR = 101.8, df = 29, $p < .001$, respectively), while the value of the LR test in the Turkish sample was nonsignificant (LR = 31.3, df = 29, $p < .35$).

The most obvious explanation for this lack of fit is the occurrence of guessing due to the multiple choice format used in the tests. In order to investigate this hypothesis the total sample was split up according to grade (e.g., the first group consisted of Dutch, Turkish, and Zambian subjects from the lowest grade tested), making a total of four groups of equal size ($N = 285$). Subsequently, an LR test was carried out on the basis of this split. The results are presented in Table 3. The LR tests revealed significant values for both instruments. More interesting however, it was observed that the relative contribution of each sample to the LR test decreased with grade. This seems to imply a poorer model fit in

TABLE 3
Contributions of Different Age Groups to the LR
Test

Group	Figure Test Log-likelihood	Letter Test Log-likelihood
1	−3850.89	−5567.76
2	−3695.79	−5409.05
3	−3289.81	−4883.85
4[a]	−3223.98	−4834.14
Total	−14296.09	−20863.01

[a] Higher group numbers refer to higher grades.

groups with lower average values, a phenomenon that may be caused by guessing, although alternative explanations, for example, modelability of test behavior as a function of performance level, cannot be ruled out.

It may be argued at this point that in view of the lack of fit to the Rasch model further analyses are not permitted on this data set. At a more abstract level this is a question of the robustness of Rasch estimates. In a simulation study by the author it has been shown that the robustness of Rasch estimates with respect to guessing depends on the criterion of evaluation. Measures derived from the difference between (either item or person) parameters and estimates proved to be heavily affected by guessing; measures of the correlation between parameters and estimates, however, were hardly influenced by guessing (van de Vijver, 1986). In the present investigation we are only interested in comparisons across samples, which are approximately equally affected by guessing. Therefore, guessing imposes no major threat here. Moreover, the correlations between the Rasch estimates of the low-scoring group and the high-scoring group turned out to be high in each cultural group. For the Figure Test these correlations were .88, .97, and .96 in the Dutch, Turkish, and Zambian sample, while values of .94, .96, and .95 were obtained for the Letter Test.

3.3. LINEAR LOGISTIC TEST MODEL

In the next phase of the analysis the Linear Logistic Test Model was applied to the data with a view to investigate the applicability of the design matrices of Figures 1 and 2: How valid is the decomposition of the Rasch item difficulty into its constituent facet levels? This analysis was first carried out at the level of the combined samples. The correlation between the Rasch estimates and the faceted item difficulties was .90. In

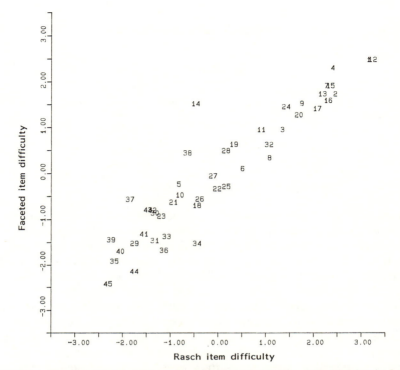

FIGURE 4. Rasch and faceted (Linear Logistic Test Model) item difficulties for the Letter Test (the digits refer to the item numbers of Figure 1).

Figure 4 the two sets of item difficulty estimates of the Letter Test are plotted. The item estimates obtained under the Rasch model were rather close to the item estimates obtained by means of facet difficulties of the Linear Logistic Test Model. In this case the correlation was .93. For both tests the item estimates seem to be decomposable into their constituent facet difficulties. Apparently, the design matrices of Figures 1 and 2 provide an effective means to define the item difficulties for the total group.

Another way of evaluating the accuracy of a design matrix involves an LR test on the basis of a comparison between the likelihood of the data matrix under the Rasch model and under the Linear Logistic Test Model. The accuracy of this test is rather questionable here, as it was found that the data matrix did not fit the Rasch model, probably because of guessing. Furthermore, guessing will probably affect the LR test of the

TABLE 4
Correlations between Original and Reconstructed Item
Parameters (Fit Test Linear Logistic Test Model)

	Figure Test	Letter Test
The Netherlands	.82	.91
Turkey	.91	.93
Zambia	.91	.92

Linear Logistic Test Model too. The power of this LR test is unknown in this case, and therefore it was not used.

After the Linear Logistic Test Model analysis of the total sample, the same analysis is repeated for each cultural group. By means of separate Linear Logistic Test Model analyses for each culture, the cross-cultural comparability of the facet difficulties can be investigated. The question is then posed whether each facet level contributes to item difficulty in the same way across cultural groups. The results of these fit tests are presented in Table 4. The correlations range from .82 to .93, indicating a good cross-cultural equivalence of the facet difficulties.

It is somewhat difficult to evaluate the size of these correlations as a clear standard of comparison is lacking. In order to investigate this a small Monte Carlo study was carried out, in which, on a random basis, a dichotomous design matrix for the Figure Test was generated, with approximately the same number of zeros and ones as in the original design matrix. Then a Linear Logistic Test Model analysis was carried out for the Figure Test on the basis of this random design matrix. After 100 analyses, each time with a different design matrix, an average correlation of .51 and a standard deviation of .11 was obtained, which demonstrates that the correlations of Table 4 are much higher than can be expected on the basis of chance.

As a rough approximation of an optimal design matrix with a particular rank, a factor analytic procedure was used. The tetrachoric correlation matrix of the Figure Test was factor-analyzed and the eight-factor solution was Varimax-rotated. Then the resulting factor loading matrix was used as the design matrix for the Linear Logistic Test Model, yielding a correlation of .97. Apparently, our design matrix of Table 4 can still somewhat be improved.

Another way of evaluating the cross-cultural comparability of the facet difficulties consists of the correlation between the Rasch parameters reconstructed by means of the estimated facet level difficulties across groups. These correlations are presented in Table 5. For both tests the

TABLE 5
Correlations between the Reconstructed
Rasch Estimates

	Dutch[a]	Turkish	Zambian
Dutch	—	.97	.95
Turkish	.97	—	.91
Zambian	.97	.98	—

[a] Figure Test above diagonal; Letter Test below
diagonal.

correlations are very high in general and typically somewhat higher for
the Letter Test than for the Figure Test. This points to the comparability
of the facet level difficulties across cultures.

3.4. Test of the Facet Hypotheses

In the final part of the data analysis the accuracy of the previously
formulated hypotheses about the facets is evaluated. The estimated facet
difficulties of the Figure Test are given in Table 6. In each cultural group
under study it was observed that items are easier when the periods of
figures repeat each other (fifth facet) or when there is a constant variation
across periods (the fourth facet), as predicted. Furthermore, it was
predicted and observed that items become easier when a larger number
of aspects is varied within a period. The same holds for the relative
difficulty of the three item-generating rules. It was found that the rule
about additions is easiest, followed by the rule about subtractions, while
the rule about alternating additions and subtractions appeared to be most
difficult.

In a final hypothesis it was postulated that items would become
easier when they dealt with a larger number of figures in each period.
The data, however, revealed a somewhat more complicated picture. The
hypothesis was only found to be adequate for items with four figures in a
period. In each group these items were the easiest ones. Items with two
or three figures were found to be equally difficult in two samples (the
Dutch and Zambian), while in the Turkish sample items with about three
figures were found to be the most difficult.

These observations can be understood when a close look at the items
of the Figure Test is taken (cf. Figure 2). The first eight figures are the
same in the five rows and the final four figures differ from one row to
another. Consequently, subjects may use the first eight figures of a row to

TABLE 6
Linear Logistic Test Model Estimates in Each Sample of
the Figure Test

Facet lebel[a]	Dutch	Turkish	Zambian
Two figures in a period	.00	.00	.00
Three figures in a period	−.01	−.41	.09
Four figures in a period	.80	.58	.54
One aspect varied	.00	.00	.00
Two aspects varied	.49	.31	.52
Three aspects varied	.66	.43	.41
Rule 1 (addition)	.00	.00	.00
Rule 2 (subtraction)	−.27	−.18	−.20
Rule 3 (alternating additions and subtractions)	−.82	−.71	−.74
Constant variation across periods	.00	.00	.00
Different variation across periods	−.36	−.54	−.86
Groups do not repeat each other	.00	.00	.00
Groups repeat each other	1.06	.93	.65

[a] First level of each facet was arbitrarily set at zero.

generate the rule applying to the item. The final four figures of a row are checked for accuracy given the rule of the item. Many subjects apparently tend to consider the ninth figure in a row as the first, that is, an arbitrary figure in a period. As long as there are two or four figures in a period, this is correct; with three figures in a period, however, the ninth figure is the third figure of the third period, that is, a nonarbitrary figure. Apparently, items with three figures in a period seem to have an additional difficulty for many subjects.

The estimated facet difficulties of the Letter Test are presented in Table 7. The results for this test can be simply summarized: all hypotheses were found to hold virtually errorlessly in each group. More specifically, items become easier as more letters are involved in the rule; this rank order is nearly perfect. Except for a small reversal in the Dutch sample, a perfect rank order was observed between the number of positions in the alphabet (the fourth and fifth item generating rule) and item difficulty, so items are more difficult when their rule applies to letters further removed from each other in the alphabet.

TABLE 7
Linear Logistic Test Model Estimates in Each Sample of
the Letter Test

Facet level[a]	Dutch	Turkish	Zambian
Rule about one letter	.00	.00	.00
Rule about two letters	1.28	.89	1.48
Rule about three letters	2.50	1.48	2.21
Rule about four letters	2.33	1.79	2.42
Rule about five letters	2.74	2.11	2.68
Rule about six letters	3.19	2.72	3.32
Rule 1	.00	.00	.00
Rule 2	.60	.86	1.32
Rule 3	.70	.65	.97
Rule 4	−1.24	−1.47	−.93
Rule 5	−1.60	−1.61	−1.14
Difference of one position	.00	.00	.00
Difference of two positions	−.72	−.29	−.16
Difference of three positions	−1.36	−.50	−.76
Difference of four positions	−1.08	−.73	−.87

[a] First level of each facet was arbitrarily set at zero.

The difficulty order of the five item-generating rules was found to be (from low to high): Rule 2 (equal letters within and across groups), Rule 3 (equal letters within periods, which are unequal across groups), Rule 1 (vowels), Rule 4 (alphabetic positions forward), Rule 5 (alphabetic positions backward).

4. CONCLUSION

The present investigation was meant as an exercise in improving our test designs. In this case Guttman's Facet Theory was adopted as an example to systematize test designs.

It was found that the theoretical structure imposed on the data could be retrieved in an empirical data set, indicating that item difficulties are rather well described in terms of the underlying facets. Even though a multiple choice answer format was used with its inherent problem of guessing, a very good fit of a Linear Logistic Test Model was found.

It should be emphasized that the test were applied in a cross-cultural context, with its almost traditional problems of item bias, which makes the present investigation a rather stringent check on the feasibility of facet designs. In analyses not reported here it was found that the classical item bias indices were low for these tests. It seems therefore that item response theory can benefit from more systematized test designs.

ACKNOWLEDGMENTS

The help of Professor C. Kagitcibasi (Bogazici University, Istanbul) and Professor R. Serpell (University of Zambia, Lusaka) is gratefully acknowledged. The data gathering in Zambia was supported by a grant from the Netherlands Foundation for the Advancement of Tropical Research (WOTRO).

5. REFERENCES

Andersen, E. B. (1973). A goodness of fit test for the Rasch model. *Psychometrika, 38,* 123–140.

Canter, D. (Ed.). (1985). *Facet theory.* New York: Springer.

Ekstrom, R. B., French, J. W., & Harman, H. H. (1976). *Kit of factor-referenced tests.* Princeton, NJ: Educational Testing Service.

Fischer, G. H. (1974). *Einführung in die Theorie psychologischer Tests.* Bern: Huber.

Fischer, G. H. (1976). Some probabilistic models for measuring change. In D. N. M. de Gruyter & L. J. Th. van der Kamp (Eds.), *Advances in psychological and educational measurement* (pp. 97–110). New York: Wiley.

Fischer, G. H. (1983). Logistic latent trait models with linear constraints. *Psychometrika, 48,* 3–26.

Hambleton, R. K., & Swaminathan, H. (1985). *Item response theory.* Boston: Kluwer-Nijhoff.

Spada, H. (1977). Logistic models of learning and thought. In H. Spada & W. F. Kempf (Eds.), *Structural models of thinking and learning* (pp. 227–262). Bern: Huber.

van de Vijver, F. J. R. (1984). *Group differences in structured tests.* Paper read at the Advanced Study Institute, Athens.

van de Vijver, F. J. R. (1986). The robustness of Rasch estimates. *Applied Psychological Measurement, 10,* 45–58.

Index

309